Nanomaterials in Bio-Medical Applications
A Novel approach

Edited by
Bichitra Nandi Ganguly

The book presents new results in the areas of nanomaterials, nanoparticles, ultra-small nanoparticles, plasmonic nanoparticles and coated nanoparticles for bio-medical applications. Emphasis is placed on (1) synthetic routes (quantum dots, thermal decomposition methods), (2) characterization methods (photo-physical techniques, X-ray diffraction, electron microscopy, light scattering, positron annihilation spectroscopy) and (3) bio-medical applications (nanomaterials and nanoparticles in physiology, medicine and bio-medicine).

Keywords

Nanomaterials, Nanoparticles, Ultra-Small Nanoparticles, Plasmonic Nanoparticles, Coated Nanoparticles, Bio-Medical Applications, Quantum Dots, Thermal Decomposition Methods, Photo-Physical Characterization, X-Ray Diffraction, Electron Microscopy, Light Scattering Characterization, Positron Annihilation Spectroscopy, Capping Ligands, Surface Ligands, Passivating Agents

Nanomaterials in Bio-Medical Applications
A Novel approach

Edited by

Bichitra Nandi Ganguly

SINP, KOLKATA, INDIA

Published by **Materials Research Forum LLC**
Millersville, PA 17551, USA

Published as part of the book series
Materials Research Foundations
Volume 33 (2018)
ISSN 2471-8890 (Print)
ISSN 2471-8904 (Online)

Print ISBN 978-1-945291-72-2
ePDF ISBN 978-1-945291-73-9

This book contains information obtained from authentic and highly regarded sources. Reasonable efforts have been made to publish reliable data and information, but the author and publisher cannot assume responsibility for the validity of all materials or the consequences of their use. The authors and publishers have attempted to trace the copyright holders of all material reproduced in this publication and apologize to copyright holders if permission to publish in this form has not been obtained. If any copyright material has not been acknowledged please write and let us know so we may rectify in any future reprint.

Distributed worldwide by

Materials Research Forum LLC
105 Springdale Lane
Millersville, PA 17551
USA
http://www.mrforum.com

Manufactured in the United States of America
10 9 8 7 6 5 4 3 2 1

Table of Contents

Preface

Chapter 1   Introduction 1

Synthetic Routes of Nanomaterials

Chapter 2   Quantum Dots, Synthesis Properties and Biology Application .. 19

Chapter 3   Synthesis of Nanoparticles through Thermal Decomposition of
Organometallic Materials and Application for Biological
Environment 50

Methodology of Physical Characterization

Chapter 4   Methodology and Physical Characterization of Nanoparticles
Using Photophysical Techniques 75

Chapter 5   Characterization of Nano-materials: X-ray diffraction method,
Electron microscopy and Light scattering 104

Chapter 6   Probing Defects by Positron Annihilation Spectroscopy 123

Bio-Medical Applications

Chapter 7   Advances in the Application of Nanomaterials and Nanoscale
Materials in Physiology or Medicine: Now and the Future 147

Chapter 8   Applications of Nano particles in Biomedicine 179

Chapter 9   Conclusive Remarks 197

Keywords 199

About the editors 200

Preface

The nanoworld opens up a wonder world for material scientists and the bio-medicine field. *Nanomaterial* is defined as a "material with any external dimension in the nanoscale or having internal structure or surface structure in the nanoscale", with *nanoscale* defined as the "length range approximately from 1-100 nanometres (nm)". This includes both *nano-objects*, which are discrete pieces of material, and *nanostructured materials*, which have internal or surface structure on the nanoscale; a nanomaterial may be a member of both these categories. Similarly, *'Nanoparticles'* are typically particles between 1-100 nm in size with a surrounding interfacial layer. The interfacial layer is an integral part of nanoscale matter, fundamentally affecting all of its properties. The interfacial layer typically consists of ions, inorganic and organic molecules. Organic molecules coating inorganic nanoparticles are known as stabilizers, capping and surface ligands, or passivating agent.

Both could be synonymous and their properties are greatly different from bulk materials. The newly emerging world in this domain also goes down to ultra small-nanoparticles and plasmonic nano particles, defining their properties which are very fascinating and drastically different from the common chemistry or physical science properties of the bulk materials. This fact exactly makes them special, as the same opens up a huge application potential.

Such properties are being utilized everywhere, because the surface atoms and the greatly enhance surface area in such particles hold many promises. Other size-dependent property changes include quantum confinement in semiconductor particles, surface plasmon resonance in some metal particles and superparamagnetism in magnetic materials. Accordingly, uses have been manifold already, for example in electro-chemistry, surface science, optics, paint industry, as catalysts in solar light harvesting agents and also in toxicology and in biomedical fields.

In this book, comprising of articles from different aspects of nano-materials and their application in bio-medical field has been highlighted, keeping in mind the surface conjugation properties of such nano-structures. The new developments and research in material science with an insight to use the nano-species in biomedicine is such an interface of the subject where the investigators face many challenges. The application of such materials in bio-medicine however requires a directed design providing actuation and stability in a particularly complex environment such as living organisms. Nanotoxicity and *in-vivo* clearances are some of limiting factors in the radiological tests such as PET/SPECT, etc.

As far as possible, the topics have been chosen with care, to suit the interest of the researchers both in material science and bio-medical field. We hope readers will find it interesting.

Bichitra Nandi Ganguly

SINP, KOLKATA, INDIA

June, 2018

Chapter 1

Introduction to Nano-Materials in Bio-Medical Applications: A Novel Approach

Bichitra Nandi Ganguly

Saha Institute of Nuclear Physics, Kolkata -700064, India

bichitra.ganguly@saha.ac.in

Nanoparticles (NPs) have evolved as novel and valuable functional building structures, notably being considered as one of the most relevant recent achievements in materials science. With an ever growing interaction between interface-research on nanoparticles and biomedicine, materials scientists are required to encounter new and exciting challenges both in the design and engineering of the material considering its targeted application as shown below in Figure 1.

A wide variety of nanoparticles are being used for active biological research and application [1-7], such as semiconductor quantum dots which not only initiate photon induced surface chemical reactions, but are also useful alternative fluorescent labelled compounds. Also, iron oxide NPs have been approved for the use in humans in magnetic resonance imaging (MRI) applications as contrast enhancers. A further novel use in current research is that of plasmonic nanoparticles due to their unique feature, i.e. displaying localized surface plasmon resonance bands in the UV-visible to near infrared spectral range [8].

Using well-established concepts and methods [9], while assessing the uniqueness of biomedical questions, nanoparticles have been used to develop ultrasensitive probes for the dreaded HIV infection and cancer, among others diseases [10,11]. New discoveries and directions using nanoparticles include theranostics and plasmonic photothermal therapy (PPT), depending up on understanding the destination of the nanoparticles. Once they are administered *in vivo*, it becomes a crucial aspect, that is under thorough investigation [12]. Core concepts of materials science that have led to novel and exciting applications in biomedicine are highlighted in this book.

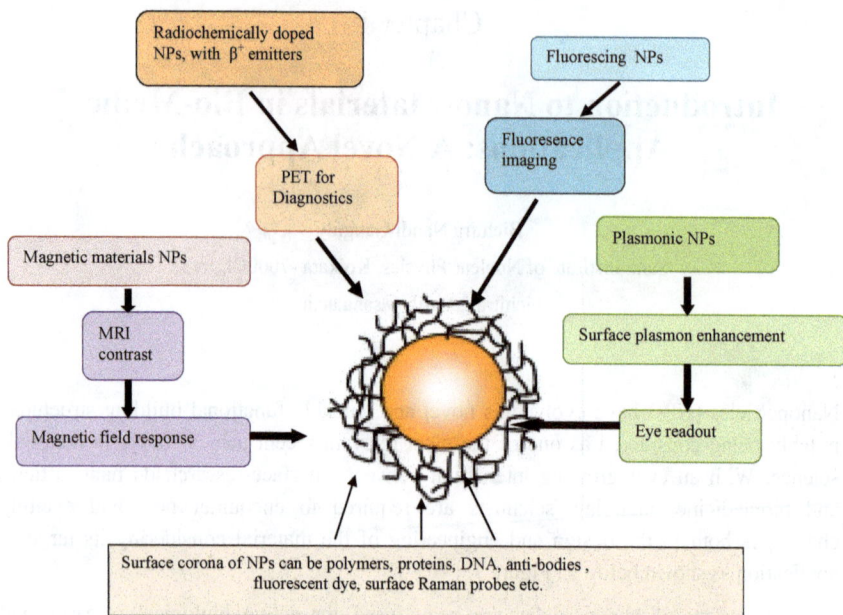

Figure 1. Functionalization, surface modification and manifold applications of nanoparticles (NPs) in biomedicine. NPs in biomedicine, are conjugated by biologically active molecules which are target specific. This new paradigm connecting nanoparticles with medical diagnostic or therapeutic use is therefore changing the point of view as well as research practices of materials scientists.

The intense interest in this specialized field (nano sized material) is due to the fact that nanotechnology involves controlled synthesis of materials where at least one dimension of the structure is less than 100 nm. This ultra-small size is comparable to naturally occurring proteins and biomolecules in the cell and is notably smaller than the typical diameter (~7 μm) of many human cells. The reduction of materials to nano-scale can frequently alter their electrical, magnetic, structural, morphological, and chemical properties enabling them to interact in unique ways with cells, biomolecules and enable their physical transport into the interior structures of cells. Nanoscale particles typically possess a larger percentage of atoms at the material's surface (see Fig. 2), which can lead to increased surface reactivity [13], and can maximize their ability to be loaded with therapeutic agents to deliver them to the target cells. As a simple illustration, one may take note of the following: in case of ZnO nano particles, from the size effect shown, the agglomeration number (n) of the molecules in the case of each samples can be explained through simple relationship (assuming the small crystallites are roughly spherical for a minimal surface to volume ratio):

$$n = 4/3\pi\, r^3\, \rho\, (N_A/M)$$

where, the density of ZnO (ρ) = 5.606 gm/cm3, N_A is the Avogadro's number, molecular weight (M) = 81.389 gm/mole as shown in Table-1.

But there can be still smaller sizes of nano particles, like ultra small nano particles (USNPs), by definition, the core size of USNPs range from 1 to 3 nm with the majority of its atoms located at the surface, both the specific surface area and the number of atoms at the surface increase drastically when the core diameter decreases towards the ultra-small range (Fig. 2). For instance, more than 70% of the atoms forming a 2-nm USNP could be located on its surface.

Table-1 A simple illustration of surface occupancy of the molecules, with the decrease in size of ZnO nanoparticles.

Average grain size, (roughly spherical)	Total no. of molecules = n	Surface /volume ratio	No. of molecules in the surface
40 nm	4×10^{24}	0.1	4×10^{23}
20 nm	2×10^{24}	0.3	6×10^{23}

This increased surface/volume ratio leads to unique properties that diverge from their microscopic species or from the bulk material itself and renders the molecules to be highly surface active. As a consequence, USNPs demonstrate the variation between small molecules and conventional, larger-sized NPs, not only in terms of size, but also as regards to their physicochemical and pharmacokinetic properties [14]. A variety of physiochemical properties such as size distribution, electrostatics, surface area, general morphology and aggregation may significantly affect physiological interactions between nanomaterials and the target biological areas.

Some nanoparticles of a particular substance are thought to pose greater risks of toxicity than larger-sized particles of the same substance [15-21]. Above all, the distribution of particles within the specimen body and the accumulation of a specific type of particle in a particular part of the body, is dependent on the particle's size and surface characteristics, that are considered critical issues.

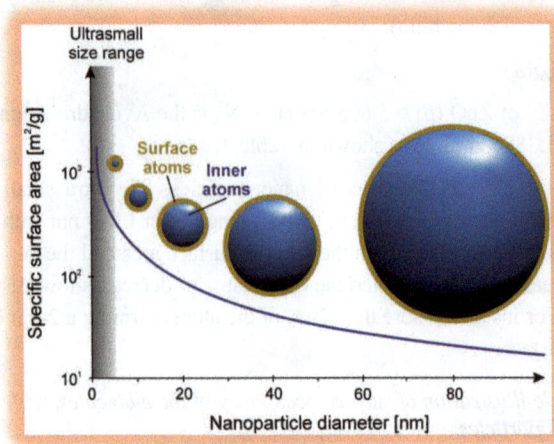

Figure 2. A simple demonstration of increase specific surface area, when the size of NP decreases, the root cause of increasing surface interaction. (ref: Nanomedicine: Nanotechnology, Biology, and Medicine 12 (2016) 1663–1701, copyright Elsevier).

Also, when the nanoparticles accumulate in specimen body system without proper excretion, it can cause continuous toxicity. The main distribution sites and target organs for nanoparticles are unknown. However, it appears that the liver and spleen may be target organs [22,23]. If nanoparticles are ingested, inhaled or absorbed through the skin, they can induce the formation of reactive oxygen species (ROS) including free radicals [24]. ROS produces oxidative stress, inflammation, and consequent damage to various biological materials such as protein, DNA, etc. Besides ROS production, other factors influencing toxicity include size, morphology, agglomeration statue, shape, chemical composition, surface structure, surface charge, aggregation and solubility [25]. As a result of their small size, nanoparticles can cross tissue junctions and even cellular membranes where they induce structural damage to the mitochondria [26,27] or invade the nucleus where they cause serious DNA mutations [28] leading to cell death [29]. The factors mentioned above can be categorized under the five characteristics of nanoparticle, which are: size; surface area; electrostatic statue of surface; morphology; and agglomeration status.

However, application and clinical administration of these particles *in vivo* requires a precondition for their rapid elimination through the system. In other words, not all USNPs can *per se* be cleared renally, as their surface charge, shape and surface composition influences their pharmacokinetics in addition to their size [30-32]. The widely used term "ultra-small nanoparticles" originating from the field of material science is therefore in no way explicably justified with the pharmacological term "renal excretable nanoparticles". In fact, their bio-distribution as well as blood clearance depends primarily on their *in vivo* hydrodynamic diameter that can be substantially larger than their *in vitro* diameter due to the unspecific adsorption of serum component including proteins and lipids. Thus, formation of a bio-molecular corona has been observed for a wide range of different NP platforms. As a result, many nano-sized objects are scavenged by the mononuclear phagocyte system (MPS) and adequate surface modifications need to be made to counter this issue and to render NPs more suitable for bio-medical applications.

The reactivity (and potential risks) posed by inorganic nanoparticles in biological environments depends on their physical and chemical state. Biological fluids such as blood, mucus, cell culture media, and others contain a large variety of substances that can interact with and modify nanoparticles. Control of interactions between nanoparticles and bio-systems is essential for the effective utilization of these materials in biomedicine. But an useful surface coronation can lead to a beneficial effect. Such as a wide variety of nanoparticle surface structures including small molecules like folic acid, giant sugars and peptides have been developed and suggested for imaging, sensing, and drug delivery applications [33-35] to cancer cells.

Nanoparticles exhibit unique physical properties, but only with proper bio-conjugation and with an appropriate hydrodynamic size, they may be accepted into bio-molecular and cellular systems. These features make them attractive materials for therapeutic and diagnostic applications (see for example Fig.3). However, utilization of these materials in biomedicine requires controlled interactions with bio-macromolecules. For example, specifically designed nanoparticle monolayer structures can impart enhanced cellular internalization ability, noncytotoxicity and improved payload binding capacity, which is therefore necessary for effective intracellular delivery. Similarly, surface functionality can be tuned to provide the selective or specific recognition required for bio-sensing. These bio-sensors are becoming increasingly important in the diagnostic field which will eventually translate into micro-devices, where actually their functional basis relies on molecular level recognition.

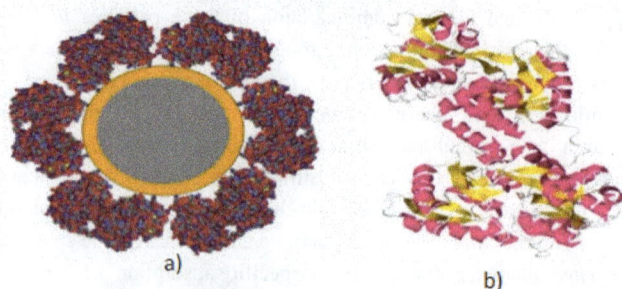

Figure 3. Nano particle with protein corona: cartoon showing: a) polymer coated nanoparticle coronated with protein molecules, b) the protein macromolecule.

As for example, biomolecules such as proteins present in biofluids, entering into contact with the NPs surface, can be adsorbed forming the so-called protein corona. There can be other molecules like giant sugar molecules or other alike soluble polymeric molecules, which may also help to perform the same function [36-38]. The bio-corona may play a key role in the *in vivo* biological fate establishing the bio-identity of the NPs. In the case of nanoparticle protein corona (Fig. 3), the outcome of the absorption of proteins onto the inorganic surface, is one of the most significant alterations. This coating provides a "biological identity" to nanoparticles in those biologically relevant fluids (e.g. commonly used cell culture medium supplemented with serum) [39,40] and determines their interaction within cells, immune systems and other components of biological systems.

As a result, detailed knowledge of the nanoparticle protein corona has emerged as a crucial aspect in understanding their bio-distribution and reactivity of nanoparticles in organisms, for the safe design of the engineered nanoparticles. Differential protein adsorption can potentially lead, for example, to different organ distribution by interacting with different tissue specific receptors [41-43].

Thus through appropriate surface conjugation, these nanomaterials can acquire the ability to selectively target particular types of cells or to pass through physiological barriers and penetrate deep into tumour sites, which has been explained below through illustrations.

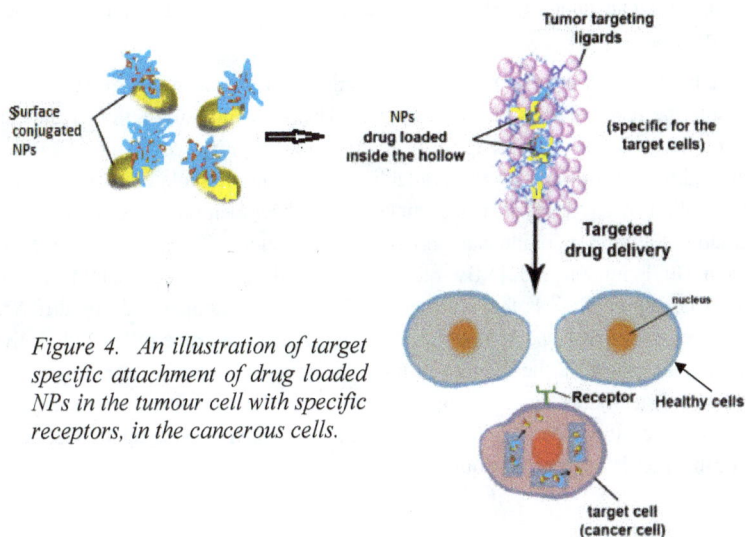

Figure 4. An illustration of target specific attachment of drug loaded NPs in the tumour cell with specific receptors, in the cancerous cells.

It is due to nano-material conjugated surface scaffold, specific pharmacokinetic properties and good tissue penetration, renal excretable USNPs or only some NPs qualify for special purposes for which other NPs with long retention time in the body are not suitable. Such classification include first and foremost their diagnostic application as molecular imaging agents since this requires fast and specific accumulation at target sites within a few hours.

Combined with rapid excretion from non-targeted tissues, this situation allows faster imaging after injection and results in high signal-to-background ratios [44,45]. semiconductor QDs [46-50] and fluorescent dye labeled USNPs [50,-52] have been developed as optical imaging probes. Additionally, some of these ultrasmall nanoparticulate probes have been utilized as intra-operative visualization tools during image-guided surgical and interventional procedures [53,54]. Depending on the attached or incorporated radiolabel, USNPs can be either applied for SPECT [52,55,56] or positron annihilation tomography/PET [57-60] imaging as found in the literature.

Single-photon emission computed tomography (SPECT) is a nuclear medicine tomographic imaging technique using gamma rays. It is very similar to conventional nuclear medicine planar imaging using a gamma camera. However, it is able to provide true 3D information.

PET on the other hand is based on recording the emission of the positrons by radioactive isotopes within the nanoparticles under study [61]. This technique offers the advantages of a definitely large sensitivity and space-resolving capability. Moreover, provided that an adequate radio-activation reaction is available, PET can in principle be used to follow the bio-distribution of any type of nanoparticles. It has been demonstrated the different bio-distribution of aluminum oxide nanoparticles, as a function of core size, in an elegant application of PET imaging [62]. By radioactive labelling of oxygen atoms in the aluminum oxide nanoparticles, the bioaccumulation of the nanoparticles in different organs could be studied *in vivo*. As noted in that study, the decay time of the radioactive isotope is relatively short, in the range of 1 hour. However, radioactive labelling cannot be used for bio-distribution studies comprising longer periods of time such as days or weeks for dynamic studies on biological pathway of the NPs, here again fast renal clearance of the drug is a foremost requirement.

Summary and conclusions

The intervention of material science in to biomedicine through the surface activity of nano materials have opened a new paradigm, which is under intensive investigation. Researchers have proposed for a variety of medical applications within the last decade, in particular for diagnosis and therapy of cancer, due to the unique properties of these materials. As it stands now, the majority of commercial nanoparticle applications in medicine are geared towards drug delivery. In biosciences, nanoparticles are replacing organic dyes in the applications that require high photo-stability as well as high multiplexing capabilities. There are some developments in directing and remotely controlling the functions of nano-probes, for example driving magnetic nanoparticles to the tumour and then making them either to release the drug load or just heating them in order to destroy the surrounding tissue. The major trend in further development of nanomaterials is to make them multifunctional and controllable by external signals or by local environment thus essentially turning them into nano-devices.

However, of late, boosting research has been devoted to USNPs as these materials display properties, such as size as well as physicochemical and pharmacokinetic characteristics, which are at the interface between molecules and larger particles. Through this introductory approach, a comprehensive overview of the current scenario of preparation, surface modification, characterization and biomedical applications of the most thoroughly investigated and emerging inorganic nano-material platforms has been outlined. Of particular importance while in application, is their potential systemic clearance via the renal pathway once they possess appropriate surface characteristics and their size is below the kidney filtration threshold. As not all USNPs can *per se* be cleared renally, the pharmacological term "renal excretable nanoparticles" can be used to recommend and to differentiate them in the future from standard NPs as too large for renal elimination. Although a tremendous amount of outstanding research has been published on this particular subset of nanomaterials, a wide variety of issues remain unexplored, unclear or unexplained.

This concerns in particular the interactions of NPs within individual cells including intracellular trafficking and precise targeting of certain cellular organelles e.g. mitochondria. As mentioned before, future research should furthermore shed light on the extent to which ligand-mediated targeting contributes to total NP accumulation at the pathological site. Here primarily a scalable, controlled and reproducible synthesis procedure resulting in defined, highly mono-disperse and uniform products under physiological conditions is an essential prerequisite. This problem commonly faced also for bigger NPs get exacerbated in the ultra-small range where even small differences in size and shape have a tremendous impact on blood circulation time, bio-distribution and elimination of NPs.

Another major issue is related to the characterization: the available techniques to characterize NPs are often not suitable for USNPs, due to instrumental limit of detection. Moreover, a lack of a precise surface characterization, once again lead to deeper consequences for NPs resulting in inconsistent outcomes *in-vivo* and *in-vitro* tests. Perfectly reproducible NPs samples are therefore a challenging goal and in order to achieve this objective, it will be necessary to radically improve the synthesis methods as well as find alternative characterization protocols and methods. Thus, many critical studies addressing the pros and cons of these materials have to be conducted in the future to deploy their full medical potential and to increase the clinical impact of *renal excretable nanodiagnostics and nanotherapeutics.* As of today, there remains immense dearth of proper and precise application of such materials through radiological processes such as theranostics, PET, MRI, etc.

References

[1]   Juan J. Giner –Casares, Malou Henriksen-Lacey, Marc Coronado-Puchau and Luis M. Liz-Marzán, Inorganic nanoparticles for biomedicine: where materials scientists meet medical research, Materials Today 19 (2016) 19-28. https://doi.org/10.1016/j.mattod.2015.07.004

[2]   Y.X.J. Wang, S.M. Hussain, G.P.Kretin, Supramagnetic iron oxide contrast agents: Physiochemical characteristics and applications in MR imaging , Eur. Radiol. 11 (2001) 2319-2331. https://doi.org/10.1007/s003300100908

[3]   O.V.Salata, Applications of nanoparticles in biology and medicine, Journal of Nanobiotechnology 2 (2004) 3-6. https://doi.org/10.1186/1477-3155-2-3

[4]   Kristof Zarschler, Louise Rocks, Nadia Licciardello, Luca Boselli, Ester Polo, P, Karina Pombo Garcia, Luisa De Cola, Holger Stephan, Kenneth A. Dawson, Kristof Zarschler, Louise Rocks, Nadia Licciardello, Luca Boselli, Ester Polo, Karina Pombo Garcia, Luisa De Cola, Holger Stephan, Kenneth A. Dawson, Ultrasmall inorganic nanoparticles: State-of-the-art and perspectives for biomedical applications, Nanomedicine: Nanotechnology, Biology, and Medicine, 12 (2016) 1663–1701. https://doi.org/10.1016/j.nano.2016.02.019

[5]   W.J. Parak, D.Gerion, T. Pellegrino, D. Zanchet, C. Micheel, C.S. Williams, R. Boudreau, M.A. Le Gros, C.A. Larabell, A.P. Alivisatos, Biological applications of colloidal nanocrystals, Nanotechnology 14(2003) R15-R27. https://doi.org/10.1088/0957-4484/14/7/201

[6]   Q.A. Pankhurst, J. Connolly, S.K. Jones, J.Dobson, Applications of magnetic nanoparticles in biomedicine, J Phys D: Appl Phys., 36 (2003) R167-R181. https://doi.org/10.1088/0022-3727/36/13/201

[7]   R Weissleder, G Elizondo, J Wittenburg, C.A. Rabito, H.H. Bengele, L. Josephson, Ultrasmall superparamagnetic iron oxide: char acterization of a new class of contrast agents for MR imaging. Radiology, 175 (1990) 489-493. https://doi.org/10.1148/radiology.175.2.2326474

[8]   M.A. Garcia, Surface plasmons in metallic nanoparticles: fundamentals and applications, J. Phys. D: Appl. Phys., 44 (2011) 283001-2833019. https://doi.org/10.1088/0022-3727/44/28/283001

[9]   M. Rubul, F.Daniel Moyano, Subinoy Rana, and Vincent M. Rotello, Surface functionalization of nanoparticles for nanomedicine, Chem Soc Rev., 41(2012) 2539–2544. https://doi.org/10.1039/c2cs15294k

[10] A.L. Martin, L.M. Bernas, B.K. Rutt, P.J. Foster, E.R. Gillies, Enhanced cell uptake of superparamagnetic iron oxide nanoparticles functionalized with dendritic guanidines., Bioconjug Chem, 19 (2008) 2375–2384. https://doi.org/10.1021/bc800209u

[11] B. Kim, G Han, B.J. Toley, C.K. Kim, V.M. Rotello, N.S. Forbes, Tuning payload delivery in tumour cylindroids using gold nanoparticles. Nat Nanotechnology, 5 (2010) 465–472. https://doi.org/10.1038/nnano.2010.58

[12] K. El-Boubbou, D.C. Zhu, C. Vasileiou, B. Borhan, D. Prosperi, W. Li, X.J. Huang, Magnetic glyco-nanoparticles: a tool to detect, differentiate, and unlock the glyco-codes of cancer via magnetic resonance imaging, Am Chem Soc.,132 (2010) 4490– 4499. https://doi.org/10.1021/ja100455c

[13] W. John Rasmussen, Ezequiel Martinez, Panagiota Louka, and Denise G. Wingett, Zinc Oxide Nanoparticles for Selective Destruction of Tumor Cells and Potential for Drug Delivery Applications, Expert Opin Drug Deliv., 7 (2010) 1063–1077. https://doi.org/10.1517/17425247.2010.502560

[14] Seung Won Shin, In Hyun Song, and Soong Ho Um, Role of Physicochemical Properties in Nanoparticle Toxicity, Nanomaterials 5 (2015) 1351-1365. https://doi.org/10.3390/nano5031351

[15] Y. Pan, S. Neuss, A. Leifert, M. Fischler, F. Wen, U Simon, G. Schmid, W. Brandau, W Jahnen-Dechent, Size-dependent cytotoxicity of gold nanoparticles, Small, 3 (2007) 1941–1949. https://doi.org/10.1002/smll.200700378

[16] D. Napierska, L.C. Thomassen, V. Rabolli, D.Lison, L. Gonzalez, M.Kirsch-Volders, J.A.Martens, P.H. Hoet, Size-dependent cytotoxicity of monodisperse silica nanoparticles in human endothelial cells., Small, 5 (2009) 846–853. https://doi.org/10.1002/smll.200800461

[17] C.Carlson, S.M. Hussain, A.M. Schrand, K. L. Braydich-Stolle, K.L. Hess, R.L. Jones, J.J. Schlager, Unique cellular interaction of silver nanoparticles: Size-dependent generation of reactive oxygen species., J. Phys. Chem. B 112 (2008), 13608–13619. https://doi.org/10.1021/jp712087m

[18] M Czajka, K. Sawicki, K Sikorska, S. Popek, M.Kruszewski, L.Kapka-Skrzypczak, Toxicity of titanium dioxide nanoparticles in central nervous system. Toxicol In Vitro, 29 (2015) 1042-1052. https://doi.org/10.1016/j.tiv.2015.04.004

[19] K.Kawata ,M. Osawa, S.Okabe, In vitro toxicity of silver nanoparticles at noncytotoxic doses to HepG2 human hepatoma cells, Environmental Science and Environ Sci Technol., 43 (2009) 6046-6051. https://doi.org/10.1021/es900754q

[20] C Carlson, S.M. Hussain, A.M. Schrand, L.K. Braydich-Stolle, K.L. Hess, R.L. Jones, J.J. Schlager, Unique cellular interaction of silver nanoparticles: size-

dependent generation of reactive oxygen species., J Phys Chem B. 112(2008)13608-136019. https://doi.org/10.1021/jp712087m

[21]   G. Oberdörster, E. Oberdörster, J.Oberdörster, Nanotoxicology: An emerging discipline evolving from studies of ultrafine particles. Environ. Health Perspect, 113 (2005) 823–839. https://doi.org/10.1289/ehp.7339

[22]   R.D. Handy, R. Owen, E. Valsami-Jones, The ecotoxicology of nanoparticles and nanomaterials: Current status, knowledge gaps, challenges, and future needs, Ecotoxicology, 17(2008) 315–325. https://doi.org/10.1007/s10646-008-0206-0

[23]   S. Hussain, K. Hess, J. Gearhart, K. Geiss, J.Schlager, In vitro toxicity of nanoparticles in BRL 3A rat liver cells. Toxicol. Vitr., 19(2005) 975–983. https://doi.org/10.1016/j.tiv.2005.06.034

[24]   J.S. Brown, K.L. Zeman, W.D. Bennett, Ultrafine particle deposition and clearance in the healthy and obstructed lung, Am. J. Respir. Crit. Care Med.,166 (2002), 1240–1247. https://doi.org/10.1164/rccm.200205-399OC

[25]   M.P. Holsapple, W.H. Farland, T.D. Landry, N.A. Monteiro-Riviere, J.M. Carter, N.J. Walker, K.V. Thomas, Research strategies for safety evaluation of nanomaterials, part ii: Toxicological and safety evaluation of nanomaterials, current challenges and data needs, Toxicol. Sci., 88 (2005) 12–17. https://doi.org/10.1093/toxsci/kfi293

[26]   A. Hoshino, K. Fujioka, T. Oku, S. Nakamura, M. Suga, Y. Yamaguchi, K. Suzuki, M. Yasuhara, K. Yamamoto, Quantum dots targeted to the assigned organelle in living cells. Microbiol. Immunol, 48(2004) 985–994. https://doi.org/10.1111/j.1348-0421.2004.tb03621.x

[27]   V. Salnikov, Y. Lukyanenko, C. Frederick, W. Lederer, V. Lukyanenko,. Probing the outer mitochondrial membrane in cardiac mitochondria with nanoparticles. Biophys., J. 2007(92) 1058–1071. https://doi.org/10.1529/biophysj.106.094318

[28]   K. Donaldson, V. Stone, Current hypotheses on the mechanisms of toxicity of ultrafine particles, Ann. Ist. Super. Sanita , 39 (2002) 405–410.

[29]   R.F. Wilson, Nanotechnology: The challenge of regulating known unknowns. J. Law Med. Ethics, 34 (2006) 704–713. https://doi.org/10.1111/j.1748-720X.2006.00090.x

[30]   L.Y. Chou, K. Zagorovsky, W.C. Chan, DNA assembly of nanoparticle superstructures for controlled biological delivery and elimination, Nat Nanotechnol, 9 (2014) 148-155. https://doi.org/10.1038/nnano.2013.309

[31]   S.Yang, S. Sun, C. Zhou, G. Hao, J. Liu, S. Ramezani, et al., Renal clearance and degradation of glutathione-coated copper nanoparticles., Bioconjug Chem., 26(2015) 511-519. https://doi.org/10.1021/acs.bioconjchem.5b00003

[32]   J.M. Tam, J.O. Tam, A. Murthy, D.R. Ingram, L.L. Ma, K. Travis, et al., Controlled assembly of biodegradable plasmonic nanoclusters for near infrared imaging and therapeutic applications., ACS Nano , 4 (2010) 2178-2184. https://doi.org/10.1021/nn9015746

[33]   Bichitra Nandi Ganguly, Buddhadeb Maity, Tapan Kumar Maity, Joydeb Manna, Modhusudan Roy, Manabendra Mukherjee, et al., L-Cysteine-Conjugated Ruthenium Hydrous Oxide Nanomaterials with Anticancer Active Application, Langmuir, 4 (2018) 1447-1456. https://doi.org/10.1021/acs.langmuir.7b01408

[34]   Bichitra Nandi Ganguly, Vivek Verma, Debanuj Chatterjee, Biswarup Satpati, Sushanta Debnath and Partha Saha, Study of Gallium Oxide Nanoparticles Conjugated with β-cyclodextrin -An Application to Combat Cancer, ACS Materials and Interfaces 8 (2016) ,17127- 17137. https://doi.org/10.1021/acsami.6b04807

[35]   Sreetama Dutta and Bichitra N Ganguly, Characterization of ZnO nano particles grown in presence of Folic Acid template, J. Nanobiotechnology 10 (2012), 29-38. https://doi.org/10.1186/1477-3155-10-29

[36]   S. Huo, H. Ma, K. Huang, Superior penetration and retention behavior of 50 nm gold nanoparticles in tumors, Cancer Res 73 (2013), 319-330. https://doi.org/10.1158/0008-5472.CAN-12-2071

[37]   J.M. Montenegro, V Grazu, A Sukhanova, S .Agarwal, JM de la Fuente, I. Nabiev, A. Greiner, W.J. Parak., Controlled antibody/(bio-) conjugation of inorganic nanoparticles for targeted delivery, Adv Drug Deliv Rev., 65 (2013), 677- 688. https://doi.org/10.1016/j.addr.2012.12.003

[38]   Diana Dehaini, H. Ronnie Fang, Liangfang Zhang, Biomimetic strategies for targeted nanoparticle delivery, Bioengineering and translational medicine, 1 (2016)   30–46.

[39]   M. Jansch, P. .Stumpf, C. Graf, E. Ruhl, R.H. Muller, Adsorption kinetics of plasma proteins on ultrasmall superparamagnetic iron oxide (USPIO) nanoparticles., Int J Pharm , 428 (2012) 125-133. https://doi.org/10.1016/j.ijpharm.2012.01.060

[40]   P. Maffre, S. Brandholt, K. Nienhaus, L. Shang, W.J. Parak, G.U. Nienhaus, Effects of surface functionalization on the adsorption of human serum albumin onto nanoparticles – a fluorescence correlation spectroscopy study, Beilstein J Nanotechnol., 5 (2014) 2036-2047. https://doi.org/10.3762/bjnano.5.212

[41]   A. Salvati, A.S. Pitek, M.P. Monopoli, K. Prapainop, F.B. Bombelli, D.R. Hristov, et al. Transferrin-functionalized nanoparticles lose their targeting capabilities when a biomolecule corona adsorbs on the surface, Nat Nanotechnol, 8 (2013) 137-143. https://doi.org/10.1038/nnano.2012.237

[42]   A. Lesniak, F. Fenaroli, M.P. Monopoli, C. Åberg, K.A. Dawson, A. Salvati, Effects of the presence or absence of a protein corona on silica nanoparticle uptake and impact on cells, ACS Nano, 6 (2012) 5845-5857. https://doi.org/10.1021/nn300223w

[43]   M.P. Monopoli, CÅberg, A. Salvati, K.A. Dawson, Biomolecular coronas provide the biological identity of nanosized materials, Nat Nanotechnol, 7(2012) 79-86. https://doi.org/10.1038/nnano.2012.207

[44]   S.C. Baetke, T .Lammers, F. Kiessling, Applications of nanoparticles for diagnosis and therapy of cancer, Br J Radiol, (2015):20150207. https://doi.org/10.1259/bjr.20150207

[45]   J.V. Frangioni, New technologies for human cancer imaging, J Clin Oncol, 26 (2008) 4012- 4021. https://doi.org/10.1200/JCO.2007.14.3065

[46]   X. Tan, R Jin, Ultrasmall metal nanoclusters for bio-related applications. Wiley Interdiscip Rev Nanomed Nanobiotechnol., 5 (2013) 569-581. https://doi.org/10.1002/wnan.1237

[47]   O. Khani, H.R. Rajabi, M.H.Yousefi, A.A. Khosravi, M.Jannesari, M Shamsipur, Synthesis and characterizations of ultra-small ZnS and Zn(1-x)Fe(x)S quantum dots in aqueous media and spectroscopic study of their interactions with bovine serum albumin, Spectrochim Acta A Mol Biomol Spectrosc .,79 (2011) 361-369. https://doi.org/10.1016/j.saa.2011.03.025

[48]   T. Xuan, S. Wang, X. Wang, J. Liu, J. Chen, H. Li, et al., Single-step noninjection synthesis of highly luminescent water soluble Cu+ doped CdS quantum dots: Application as bio-imaging agents, Chem Commun (Camb), 49 (2013) 9045-9047. https://doi.org/10.1039/c3cc44601h

[49]   L.N. Chen, J. Wang, W.T. Li, H.Y. Han, Aqueous one-pot synthesis of bright and ultrasmall CdTe/CdS near-infrared-emitting quantum dots and their application for

tumor targeting in vivo, Chem Commun (Camb), 48 (2012) 4971-4973. https://doi.org/10.1039/c2cc31259j

[50]   Y. Li, Z. Li, X. Wang, F. Liu, Y. Cheng, B. Zhang, et al., In vivo cancer targeting and imaging-guided surgery with near infrared-emitting quantum dot bioconjugates, Theranostics, 2 (2012) 769-776. https://doi.org/10.7150/thno.4690

[51]   K. Ma, H. Sai, U. Wiesner, Ultrasmall sub-10 nm near-infrared fluorescent mesoporous silica nanoparticles, J Am Chem Soc, 134 (2012), 13180-131803. https://doi.org/10.1021/ja3049783

[52]   C. .Zhou, G. Hao, P. Thomas, J. Liu, M. Yu, S. Sun, et al., Near-infrared emitting radioactive gold nanoparticles with molecular pharmacokinetics, Angew Chem Int Ed Engl, 51 (2012) 10118- 10122.

[53]   A. Mignot, C. Truillet, F. Lux, L. Sancey, C. Louis, F. Denat, et al., A top down synthesis route to ultrasmall multifunctional Gd-based silica nanoparticles for theranostic applications, Chemistry, 19 (2013) 6122-6136. https://doi.org/10.1002/chem.201203003

[54]   M.S. Bradbury, E. Phillips, P.H. Montero, S.M. Cheal, H. Stambuk, J.C. Durack, et al., Clinically-translated silica nanoparticles as dual-modality cancer-targeted probes for image-guided surgery and interventions, Integr Biol (Camb), 5 (2013) 54-86. https://doi.org/10.1039/C2IB20174G

[55]   Kotb Shady, Alexandre Detappe, Florence Appaix, L. Emmanuel. Barbier, Vu-Long Tran, Marie Plissonneau, Hélène Gehan, Florence Lefranc, Claire Rodriguez-Lafrasse, Camille Verry, Ross Berbeco, Olivier Tillement, and Lucie Sancey, Gadolinium-Based Nanoparticles and Radiation Therapy for Multiple Brain Melanoma Metastases: Proof of Concept before Phase I Trial, Theranostics, 6 (2016) 418–427. https://doi.org/10.7150/thno.14018

[56]   Y. Li, Z. Li, X. Wang, F. Liu, Y. Cheng, B. Zhang, et al., In vivo cancer targeting and imaging-guided surgery with near infrared-emitting quantum dot bioconjugates. Theranostics, 2 (2012) 769-776. https://doi.org/10.7150/thno.4690

[57]   Y. Zhao, D. Sultan, L. Detering, H. Luehmann, Y. Liu, Facile synthesis, pharmacokinetic and systemic clearance evaluation, and positron emission tomography cancer imaging of ^{64}Cu-Au alloy nanoclusters., Nanoscale, 6 (2014) 13501-13509. https://doi.org/10.1039/C4NR04569F

[58]   F. Gao, P Cai, W.Yang, J. Xue, L. Gao, R. Liu, et al., Ultrasmall [^{64}Cu]Cu nanoclusters for targeting orthotopic lung tumors using accurate positron emission

tomography imaging., ACS Nano, 9 (2015) 4976-4986.
https://doi.org/10.1021/nn507130k

[59] M. Zhou, J. Li, S. Liang, A.K. Sood, D. Liang, C. Li,. CuS nanodots with ultrahigh
efficient renal clearance for positron emission tomography imaging and image-
guided photothermal therapy, ACS Nano, 9 (2015) 7085-96.
https://doi.org/10.1021/acsnano.5b02635

[60] R. Chakravarty, S. Goel, A. Dash, W. Cai, Radiolabeled inorganic nanoparticles
for positron emission tomography imaging of cancer: an overview, Q J Nucl Med
Mol Imaging, 61 (2017))181-204.

[61] B.N. Ganguly, N.N. Mondal, M. Nandy, F. Roesch, Some physical aspects of
positron annihilation tomography: A critical review, Journal of Radioanalytical
and Nuclear Chemistry, 279 (2009) 685–698. https://doi.org/10.1007/s10967-007-
7256-2

[62] C. Pêrez-Campana, Vanessa Gómez-Vallejo, Maria Puigivila, Abraham
Martín, Teresa Calv, Carlos Pérez-Campañao-Fernández, Sergio E. Moya, Ronald
F. Ziolo, Torsten Reese, and Jordi Llop, Biodistribution of Different Sized
Nanoparticles Assessed by Positron Emission Tomography: A General Strategy
for Direct Activation of Metal Oxide Particles, ACS Nano, 7(2013) 3498-350.
https://doi.org/10.1021/nn400450p

Synthetic Routes of Nanomaterials

Nanoparticle synthesis refers to methods for creating nanoparticles. Nanoparticles can be derived from larger molecules, or synthesized by 'bottom-up' methods that, for example, nucleate and grow particles from fine molecular distributions in liquid or vapour phase. Synthesis can also include functionalization by conjugation to bioactive molecules.

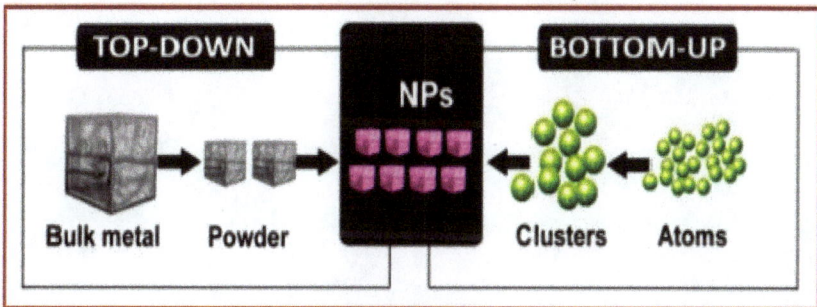

Chapter 2

Quantum Dots, Synthesis Properties and Biology Application

Nimai Mishra

Department of Nanochemistry, Italian Institute of Technology, GenovaVia Morego, 30, 16163 Genova, Italy

nimai.mishra@iit.it, nimaiiitm@gmail.com

Abstract

Quantum dots (QDs) are semiconductor nanocrystals which show extraordinary optical and electrical properties due to quantum confined nature of their energy levels. For a given material, the variation in optical properties of the QD stem from its size. Typically, the size ranges from a few nanometers to tens of nanometers. Such small dimensions are usually smaller than the de- Broglie wavelength of thermal electrons. Due to quantum confinement effect these particles exhibit unique optical properties, such as narrow FWHM, high quantum yield, size dependent emission etc. which make them useful for the several biological applications such in bio-imaging and diagnosis. In this chapter, we describe the synthesis of different QDs, their functionalization, processing and their combination with suitable materials to match the needs of biological applications.

Keywords

Semiconductor Nanocrystals, Quantum Dots, Synthesis, Optical Properties, Bio-imaging

Contents

1. Introduction...20

2. Colloidal Synthesis of mono-dispersed semiconductor nanocrystals .21

3. Size dependent properties of semiconductor nanocrystals..................23

4. Surface structure of the quantum dots.......................................24

5. Quantum confinement effects in semiconductor nanocrystals 27

6. Doped semiconductor nanocrystals 31

7.   Water-soluble colloidal nanocrystals 32

8.   QDs in biological imaging 33

9.   *In-vitro* imaging 33

10.   *In-vivo* imaging 35

11.   Toxicity 36

12.   Conclusion 37

8.   References 37

1.   Introduction

In this chapter semiconductor nanocrystals (NCs), their size dependent optical properties and surface functionalization for bio-application will be discussed. These nanocrystals are tiny and consist of hundreds to thousands of atoms [1-3], sometimes they are of sub-nanometer sizes, comprise of a new class of materials and have been a hot topic in research for the past two decades because of their unique size dependent properties. Due to the quantum confinement effect, the electronic structure and optical properties of semiconductors can be tuned by varying the physical size of the crystal to obtain new properties. Other advantages of these NCs are their thermodynamic stability and are easy to handle when present in colloidal solutions. These unique properties open new avenues for device applications. Several optoelectronic applications such as luminescent tags [4], light emitting devices as well as lasers [5-7], light-emitting diodes (LEDs) [8], solar cells [9], biological applications [10-14], etc. have been realized because of the size dependent properties of these semiconductor NCs stemming from the quantum confinement effect. In this chapter, we will discuss colloidal synthetic techniques of these semiconductor nanocrystals. Next we will show the size dependent properties of colloidal nanocrystals. Among the variety of different materials, NCs composed of semiconductor, commonly known as quantum dots (QDs), are particularly interesting because of their quantum confinement properties and excellent size dependent optical properties. The different size dependent and surface properties of quantum dots are discussed, which includes surface functionalization for biological application, blinking properties and doping. At the end we will summarize the use of QDs in bio-imagining which includes *in vitro* and *in vivo* use.

2.   Colloidal synthesis of mono-dispersed semiconductor nanocrystals

With the help of solution-phase colloidal chemical process, uniform shaped nanocrystals have been successfully synthesized over the last two decades. The different routes to synthesize these NCs are: reduction [15-17], the sol–gel process [18,19], and hot injection thermal decomposition [20,21]. Among these, the hot injection methods will be elaborated in details due to its wide applicability to produce highly crystalline and mono-dispersed nanocrystals.

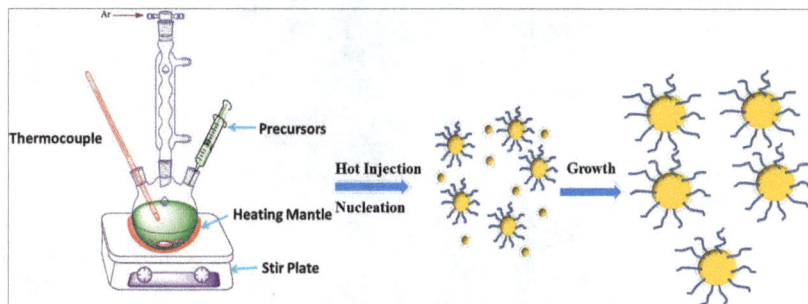

Figure 1. Hot injection method for the synthesis of mono-disperse nanoparticles.

The thermal decomposition methods or hot injection methods, where organometallic precursors of the metal–surfactant complexes have been rapidly injected into a mixture of degassed, high boiling-point organic solvent and surfactants at high temperatures under inert conditions. In 1993, the Bawendi group at MIT introduced the hot injection method (as depicted in Fig. 1) to produce mono-disperse spherical and highly crystalline cadmium chalcogenides [20]. In this method, organometallic precursors such as dimethyl cadmium and trioctyl-phosphine-selenide (TOP-Se) injected into hot (>360°C) coordinating solvents induces a burst of nucleation, and the subsequent growth generates mono-dispersed CdSe nanocrystals. By varying the experimental condition, the size of the nanocrystal could be tuned from 1.2 nm to 12 nm and they exhibit quantum confinement effect (this will be discussed in more details in the next section). This high temperature hot injection, organic-phase synthetic method has been used widely to synthesize nanocrystals because of its many advantages, such as mono-dispersity, the high crystallinity and their stability in organic solvents. The ligands (also known as capping groups) control the nucleation and growth rates by dynamically binding to and coming off of the surface of the nanocrystals, as well as to the constituent precursors in solution. These surface capping groups enable control over the size and shape of the nanocrystals. Due to these synthetic advantages over other methods, this pioneering work by the Bawendi group has been widely extended to the synthesis of nanocrystals of various materials such as semiconductors, metals, and metal oxides.

Following the hot injection methodology reported by the Bawendi group, the synthetic development of various II-VI (CdSe,CdTe, CdS) [22-27], III-V (InP,InAs) [20-31] and IV-VI (PbS [32], PbSe [33], PbTe [34]) colloidal semiconductor QDs have since been reported.

Figure 2. TEM images of colloidal semiconductor nanocrystals of different materials. Adapted with permission from reference 35.

Over the years, significant progress has been made in the synthesis of different technologically important semiconductor nanocrystals. Fig. 2. shows the TEM images of monodispersed nanocrystals of some of those materials. These syntheses are carried out at high temperatures, and in the presence of long-chain alkyl-phosphines (e.g. trioctylphosphine, TOP), alkyl- phosphineoxides (e.g. trioctylphosphine oxide, TOPO), alkylamines (e.g. hexadecylamine, HDA), and alkyl-phosphonic acids as surface capping groups as well as coordinating solvent. In the case of the original hot-injection method introduced by Murray et al. in 1993, quantum yields (QYs) were often low: roughly ~1 – 5%. This problem can be overcome by the growth of another inorganic material of a larger band gap around the semiconductor nanocrystals (core-shell), as first reported by Hines et al. in 1996 [36]. In this process, the shell material precursors are added drop wise to a relatively dilute solution of semiconductor nanocrystal cores at temperatures sufficiently low to prevent homogeneous nucleation of the shell materials. This requires that when the shell material is being chosen, the band gap and lattice mismatch should be taken into consideration. In the ideal case of epitaxial growth of the shell around the semiconductor nanocrystals, the dangling bonds become well-passivated, thus reducing the number of non-radiative recombination processes and dramatically improves the overall QY of the semiconductor nanocrystals. In fact, it was recently shown that spherical core-shell CdSe/CdS semiconductor nanocrystals can have QYs of nearly 100% [37].

In the next section, we will discuss the surface structure of the semiconductor nanocrystals and their effects towards shape dependent optical properties. Fig. 3 demonstrates several examples of metal nanocrystals synthesized by similar hot injection based colloidal chemistry. The reduction of metal ions by using reducing agent such as boron- hydride, amines, or 1,2-diols in the presence of surface capping groups, typically long-chain alkane-thiols: e.g. dodecanethiol, amines, or fatty acids is one of the most viable path to synthesize several metal nanocrystals. This methods works quite well for noble metals (Au [38,39], Ag [40,41], Pt [42,43], Pd [44]) and their alloys (e.g., Au-Ag [45]).

3. Size dependent properties of semiconductor nanocrystals

Colloidal semiconductor nanocrystals, also known as "quantum dots" (QDs), are inorganic materials composed of atoms ranging from a few hundred to a few thousand in number, capped by an organic layer of surfactant molecules known as ligands [46]. Their small size (radii within 1-10 nm) leads to quantum confinement effects, which causes the energy levels near the band edge to become discrete (Fig. 3). The photograph below shows the shape-dependent emission of CdSe QDs of different sizes as visible to the naked eye under UV illumination, where the smallest particles emit blue and the largest particles emit red. This quantum confinement effect normally occurs when the particle size is comparable to its bulk Bohr exciton radius (an example: 5.6 nm for the CdSe).

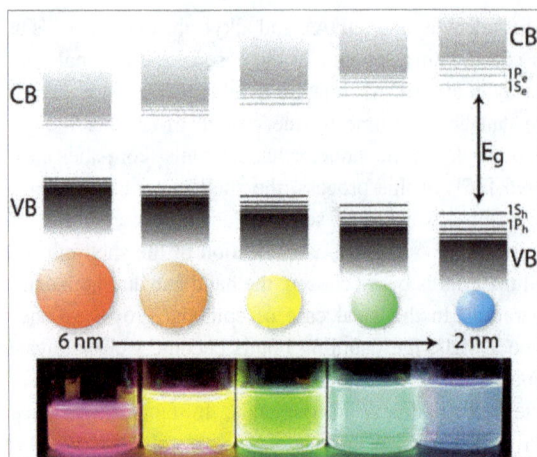

Figure 3. Schematic representation of the quantum confinement effect on the energy level in semiconductor nanocrystals. The lower panel shows colloidal suspensions of CdSe NCs of different sizes under UV excitation. Adapted with permission from reference 47.

4. Surface structure of the quantum dots

Due to the high surface-to-volume ratio in QDs, optical properties of QDs are highly dependent on the electronic quantum states associated with the surface, the so called surface states. This dependency can be better understood by the fact that, for the case of 5 nm diameter CdS, roughly 15% of the atoms are on the surface [48]. Such a high density of surface sites can either facilitate a pronounced or reduced transfer rate of photo-generated charge carriers to surface states. These states may cause several effects on various properties of the QDs, such as quantum efficiency, spectral profile, and aging [49]. The energies of these surface states generally lie within the band-gap of the QDs [50]. Therefore, the surface states can trap charge carriers (electron or hole) and function as reducing (electron-donating) or oxidizing (hole-donating) agents. Given their relatively long lifetimes, charge carriers trapped within surface states can significantly affect the overall conductivity and optical properties of QDs. Thus, the study of optoelectronic properties of QDs very much concerns understanding the nature of the surface states and how to prevent their occurrence. As it is shown in Fig. 4A-C, nanocrystals are highly faceted and each surface contains a periodic array of unpassivated orbitals (dangling bonds) which form a band structure.

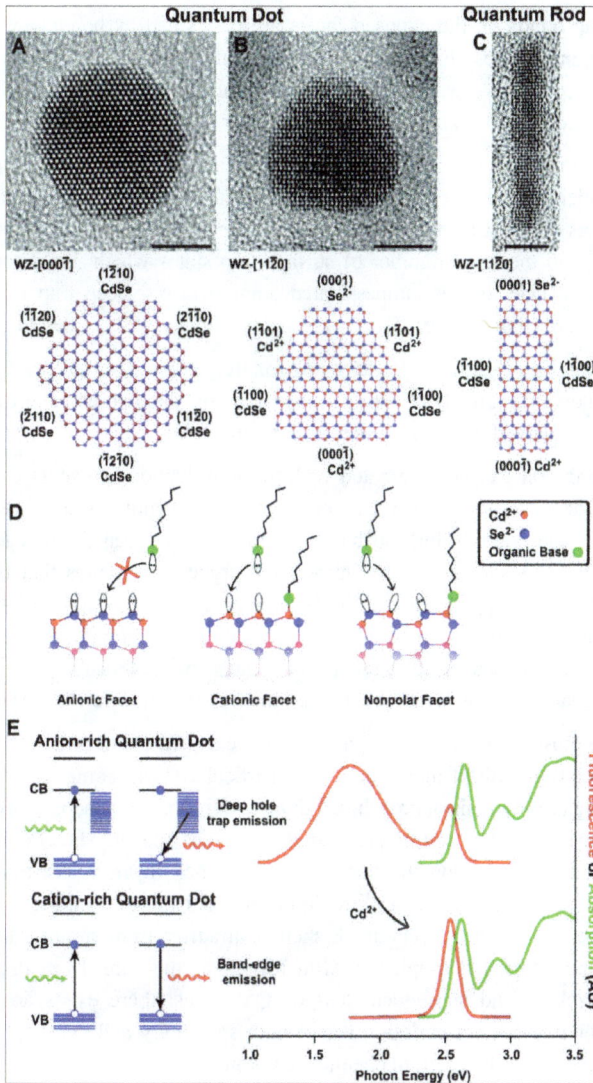

Figure 4. Surface properties of CdSe nanocrystals. (A-C) TEM images of CdSe quantum dots and rods with their atomic model. All scale bars are 5 nm.

The dangling bonds on the exposed facets are passivated by bonding with atoms from surface capping groups (Fig. 4D). For most of the cases basic moiety of the surface capping groups satisfy the cation rich and neutral facets through dative bonding while leaving anionic facets unsatisfied (Fig. 4D).

In the atomic models, the crystalline orientations and lattice facets (cation or anion rich facets) are identified by their wurtzite (WZ) structure. It is often found that, nanocrystals with surfaces terminated mostly by anions typically have little or no fluorescence emission due to the large number of surface trap states which facilitates non-radiative recombination. Due to these unpassivated anionic facets, deep trap emission is often observed in nanocrystals (Fig. 4E).

Miller indices. (d) Describes the terminal dangling orbitals on each type of facet, and (Fig. 4E) shows the effects of surface hole traps on the fluorescence of small 2.1 nm nanocrystals. Adapted with permission from reference 51.

This trap emission can be eliminated with the introduction of excess Cd^{2+} ions, to the anion rich nanocrystals, which can passivate the selenium states, yielding cationic surfaces that can strongly bind to basic ligands. As has been discussed earlier, mono-dispersed QDs are synthesized by introducing organic molecules that bind to the QD surface and act as capping agents. These organic capping groups are used not only to allow for dispersibility in compatible solvents, but also to provide a means to conjugate them to other molecules relevant to biological or chemical sensing applications. However, organic ligands are unable to bind to all of the atoms at the surface of the QD.

Perhaps the most effective way to passivate the exposed atoms at the QD surface is the use of epitaxial growth of inorganic layers, particularly with a material that possesses a larger band-gap [36]. This approach can dramatically improve the QY because all of the surface atoms of the QD can be passivated by the shell layer. The QY of the core/shell QD is also dependent upon the thickness of the shell layer, where a very thick shell ensures that the photogenerated exciton in the core is isolated from the external chemical environment. Nevertheless, very thick shells can suffer from strains caused by lattice-mismatch and become non-epitaxial with respect to the core. Even thicker shells can result in cracking and subsequent loss of QY. Thus, there exists an optimum shell thickness for maintaining emission wavelength specificity and QY while being able to isolate the core from the external chemical environment.

Depending on the band alignment between the valence and conduction bands of the two adjoining materials in core/shell structures, different charge carrier localization regimes can be observed after photoexcitation. There are three possible charge carrier localization regimes: Type-I, Type-I$^{1/2}$ and Type-II [52]. In the case of Type-I band alignments, the band gap of the core semiconductor lies entirely within the gap of the other material and upon photoexcitation the exciton is mainly confined in the core material. In the case of Type-II structures the band edges form staggered energy level alignment and results in the spatial separation of e⁻ and h⁺ on different sides of the hetero-junction, leading to the formation of a longer lived exciton, which is beneficial for photodetectors and photovoltaic applications. In the remaining case, Type-I$^{1/2}$ structures feature a direct alignment of one of the band edges but a misalignment in the other (also known as "quasi type-II"), one carrier is confined in the core or the shell, while the other carrier is delocalized over the whole core/shell structure. The QY is improved due to a second inorganic shell on to QDs core, but it still has organic capping ligand on the core/shell QDs surface which can hinder the electronic application as those organic group possess high energy barrier.

In recent years, Talapin's group developed a new approach to overcome this problem where they replaced the original long-chain organic ligand with soluble molecular metal chalcogenide complexes (MCCs) such as Zintl ions (e.g. SnS_4^{4-}, $Sn_2Se_6^{4-}$, $In_2Se_4^{2-}$, $Ge_4S_{10}^{4-}$) and one-dimensional metal chalcogenide chains solvated by hydrazinium cations and/or neutral hydrazine molecules (e.g., $(N_2H_4)_2ZnTe$) [53].

5. Quantum confinement effects in semiconductor nanocrystals

For semiconductors, the band-gap is the energy required to excite an electron from the valence band to the conduction band, leaving a hole in the valence band (Fig. 5A). In the absence of an external field, photo-excitation of the semiconductor results in a bound electron-hole pair, called an exciton. The exciton behaves like a hydrogen atom, except that a hole and not a proton is present. The mass of a hole is much smaller than that of a proton, and this property of the hole affects the solutions to the Schrödinger wave equation. The distance between the electron and the hole is called the exciton Bohr radius (R_b). If the radius (R) of a QD approaches R_b, i.e., $R \approx R_b$, or $R < R_b$, the motion of the electrons and holes become confined spatially to the dimensions of the QD and this causes an increase of the excitonic transition energy and the observed blue shift in the QD band-gap luminescence. This is the origin of the so-called quantum confinement effect. Discretization of the electronic states for semiconductor QDs (illustrated in Fig. 5A) can qualitatively be understood by considering a "particle-in-a-box" model, where the electron and hole start to strongly "feel" the effects of the boundary which is at a higher energy potential.

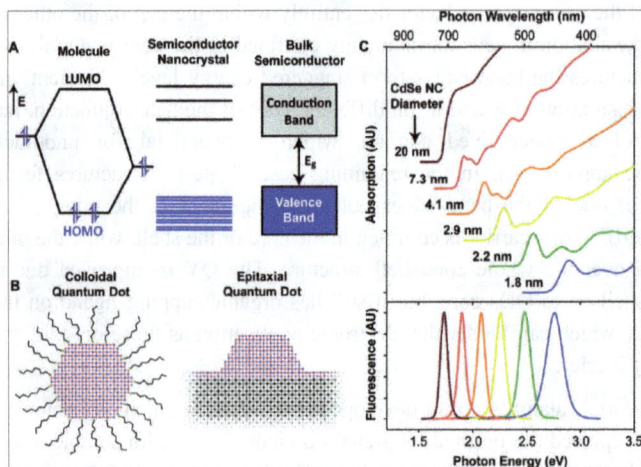

Figure 5.(A) Electronic energy states of a semiconductor in the transition from discrete molecules to nanosized crystals and bulk crystals. (B) Comparison of a colloidal quantum dot and epitaxial grown quantum dots on a crystalline substrate. (C) Absorption (upper) and fluorescence (lower) spectra of CdSe semiconductor nanocrystals showing quantum confinement and size tunability. (AU)arbitrary units. Adapted with permission from reference.51.

In general, quantum confinement effects become important when the particle radius is comparable to or smaller than the Bohr exciton radius. It is known that the bulk Bohr exciton radius of CdSe is 5.6 nm. As seen in the reported absorption spectra given in Fig.5C, below, the confinement effect is clearly seen for QDs below 5.6 nm in radius (as shown by the multiple distinct peaks which correspond to discrete energy transitions) and is much less pronounced for QDs larger than that.

Upon excitation from an external energy source, an electron from the valence band is excited to the higher energy conduction band and this creates an exciton (i.e. electron-hole pair). After the formation of an exciton, the electron may recombine radiatively with the hole and relax to a lower energy state, ultimately reaching the ground state (Fig. 5C).

The most common radiative relaxation processes in intrinsic semiconductors is band-edge and near band-edge (exciton) emission. The energy difference between the maxima of the emission band and of the maxima of the lowest energy absorption band is called the Stokes shift. The electron and hole are bound by a few m eV to form an exciton, and radiative recombination of that exciton results in near band-edge emission at energies slightly lower than the band-gap due to the Stokes shift. The lowest energy states in QDs are referred as 1_{se}-1_{sh} (also called exciton state). The additional peaks along with 1_{se}-1_{sh} were observed in QD absorbance spectra, Bawendi et al. assigned these peaks as formally forbidden 1_{se}-1_{ph} and 1_{se}-2_{sh}. The full width at half maximum (FWHM) of a room-temperature band-edge emission peak from QDs varies from 15 to 30 nm depending on the degree of polydispersity in the QDs, average size of the QDs. etc. Occasionally radiative emission occurs due to presence of defects states. Defect states are within the QDs bandgap [54]. Due to the nature of the defect or impurity it can act as a donor (has excess electrons) or an acceptor (has a deficit of electrons). Columbic attraction draws electrons or holes to these sites of deficient or excess local charge. These defect states can be categorized into either shallow or deep levels, where shallow level defect states have energies near the conduction band or valence band-edge. In most cases, shallow defects exhibit radiative relaxation at temperatures sufficiently low so that thermal energies (kT) do not excite the carriers out of the defect or trap states. On the other hand, deep levels are so long-lived that they typically experience nonradiative recombination. The efficiency of these radiative process is very important for display application. This can be quantified via quantum yield measurements which calculate number photons emitted by these QDs upon excitation with known number of photons.

One procedure to determine the QY is to simply compare the integrated emission intensity from the QDs to that of other emissive standards. The optical densities of the QDs and the standard/s are determined. The absorbance value (units: optical density – O.D.) for both samples and standards should be kept below 0.08 at the excitation wavelength.

Instead of radiative recombination of excited electrons and holes, there is an alternative process by which excited states recombine to lower energy and ground states through non-radiative paths, and is termed non-radiative recombination. The non-radiative relaxation may be categorized as two types: first, internal conversion and external conversion, and second, Auger recombination [55]. Nonradiative recombination via local strain in a lattice that can create a local potential well which can also trap electrons and holes, which is a common phenomenon in internal conversion. Another important process is Auger nonradiative recombination, where primary strong carrier-to-carrier interaction is responsible. In the Auger process, rather than releasing the energy of recombination as a photon or phonon, the excess energy is transferred to another electron and involves two electrons and a hole in the conduction and valence bands, respectively.

Apart from these nonradiative procesees "blinking" phenomena is one of the main hurdles towards practical utilization of these QDs in light emitting devices across a wide spectral range. "Blinking" is a random switching process between bright and dark states, that is emissive and non-emissive states and this is observed in fluorescent colloidal QDs when examined at a single particle level [56].

Recently, several reports suggested a potential solution to the problem of blinking: nanocrystal shell engineering to control recombination paths, leading to almost non-blinking nanocrystals [57]. There are several mechanisms attributed with this unusal blinking behaviour and among them, the 'Auger model' has been most widely accepted [55,56]. In the 'Auger model', the bright or "on" state corresponds to a charge-neutral QD, while the dark or "off" condition occurs when the QD carries an extra charge. When a trion configuration forms in the nanocrystal (one exciton plus another charge) [58-60], this extra charge can kill the radiative process by accepting energy from the trion in a non-radiative Auger recombination process. To overcome this problem, researchers have recently developed a unique but rather simple synthetic approach where a secondary thicker shell was grown on to the QDs. Thick shell QDs, often termed as "giant" QDs, are a novel type of core/shell quantum dot that exhibit almost 100% blinking suppression, or non blinking emission, which also shows strongly suppressed Auger recombination. Initial work from the Hollingsworth group yielded suppressed-blinking in CdSe/CdS QDs with ultra-thick shell or "giant" CdSe/CdSNQDs (g-NQDs) [57]. In this case, the shell was prepared by successive addition of almost 20 monolayers of shell material.

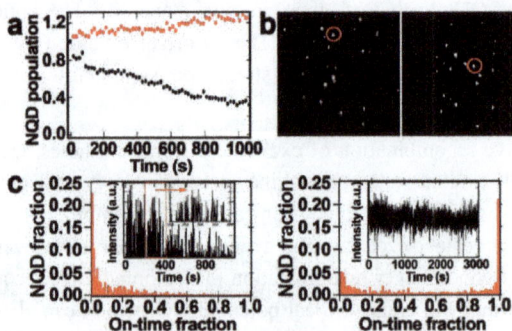

Figure 6. (a) Emitting fraction over time shows stable for giant core/shell QDs (red line) and photo bleaching for normal dots (only CdSe cores): (b) Showing fluorescence image with temporal resolution of 20 ms and (c) On-time fraction histograms of normal QDs (left) and then giant core/shell QDs (right). Insets show blinking traces for both normal and giant QDs (b). Adapted with permission from reference. 57.

After a detailed investigation of these giant QDs on the single-dot level, ensemble-level performance was found with enhanced photo stability. Under continuous excitation, long, ~1-hour blinking statistics for a large population of dots were measured and it was found that >20% of the g- NQDs were entirely non blinking (on-time fraction >0.99) and >40% showed significantly suppressed blinking (on-time fraction>0.8) (Fig. 6).

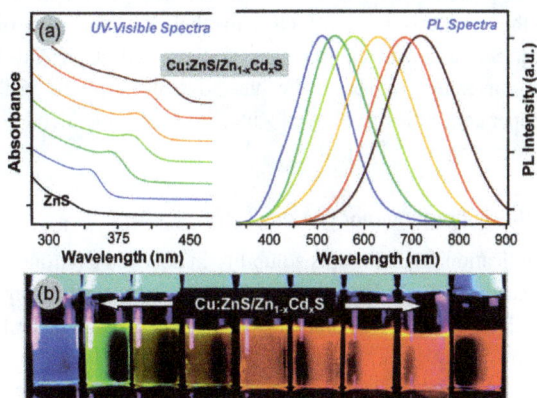

Figure 7. (a) Evolution of UV-visible (left panel) and corresponding photoluminescence (right panel) during doping of Cu in the alloyed nanocrystals. (b) Digital picture of samples under UV light, collected in a typical experiment from different stages of the doping process of the alloyed nanocrystals. Adapted with permission from reference 70.

6. Doped semiconductor nanocrystals

In recent years, much effort has been devoted into the research of impurity atoms doped semiconductor nanocrystals. Studies have found that impurities like Mn, Cu, Co, etc., doped into semiconductor nanocrystals could strongly influence the electronic, optical and magnetic properties [61-63]. The emission can be tuned across the whole visible range up to near infrared and this can be achieved either by changing the dopant or host material. One of the earliest reports of doping in semiconductor nanocrystal system was from Bhargava et al. [64]. Several approaches have been developed to synthesize high quality doped semiconductor nanocrystals. Here we describe three most studied methods for doping in semiconductor nanocrystals system. In the first synthesis process, where doping occurs inside the host material during the growth process, there is a clear increase in the intensity of dopant emission suggesting a larger fraction of nanocrystals being doped. In the second synthesis technique, named nucleation doping, dopant chalcogenide clusters are first formed, and host semiconductors are grown on them, as can be seen in Fig. 7. In the third synthesis process, the dopant (e.g. Mn) is incorporated into the shell layer (ZnS) on top of a core of nanocrystals (CdS).

The properties of doped nanocrystals are very strongly linked to the doped atomic species and content. Different kinds of metal atoms such as, copper, cobalt and silver have been introduced into the synthesis of II–VI QDs to alter the properties of semiconductor materials [65-70]. Srivastava et al. reported efficient and stable Cu-doped QDs with tunable emission wavelength covering the blue to red end of the visible spectrum by incorporating copper impurities to alloyed $ZnS/Zn_{1-x}Cd_xS$ zinc-blend (ZB) nanocrystals as it has been shown in figure 7 [71].

7.   Water-soluble colloidal nanocrystals

For biological application quantum dots solubility in aqueous environment is needed, for that nanocrystal surface should be coated with hydrophilic surface capping groups. To date different kinds of quantum dots such as CdS [72-79] and CdTe [80,81] has been synthesized directly in aqueous medium, which is different than the above mentioned colloidal hot injection methods. The main limitation of aqueous medium synthesized quantum dots is theat they are poor in fluorescent quantum yield and size distribution, but at the same time for biological application, a very good water solubility is needed. Therefore, researchers mainly synthesized QDs in organic, high temperature synthesis medium and thereafter transfer them in aqueous medium for biological application. There are several strategies that have been developed to transfer those conventional quantum dots into water. The first one which is pioneered by the Nie and Weiss groups is making of water-dispersed QDs by ligand exchange method, which can be seen in Fig. 8 [82,83]. The other approach is coating of polymer and lipid on to the QDs which can efficiently make this material soluble in water [84,85]. The easiest way to obtain a hydrophilic surface is by exchanging the hydrophobic fatty acid (oleic acid), alkyl amine, TOPO, TOP surfactant molecules with bi-functional molecules that are hydrophilic on one end and the other end binds to the QD surface. Commonly, a bi-functional ligand such as mercaptocarbonic acid (HS---------COOH), 11-mercaptodecanoic acid thiols (–SH) are used that can bind to the QD surface and carboxyl (–COOH) groups are a prominent example for hydrophilic groups [86-91]. On the QDs surface the negatively charged carboxyl groups (at neutral pH) repel each other electrostatically. Due to this in the electrolytic solution the QDs precipitate at salt concentrations of a few hundred millimolar [92] concentration. It is because of this reason, the water solubility of QDs capped with mercaptocarbonic acids is therefore limited. The stability in aqueous medium can be enhanced by using mercaptocarbonic acids with two instead of just one thiol group [93]. Another possibility for stabilization of QDs nanocrystals in aqueous solutions might be to cover them with either protein molecules, polymer molecules that can be adsorbed on the surface of nanocrystals [94]. Yet another more stable route is silanization on to the surface of QDs, which is growing a glass shell around the particles [95-98].

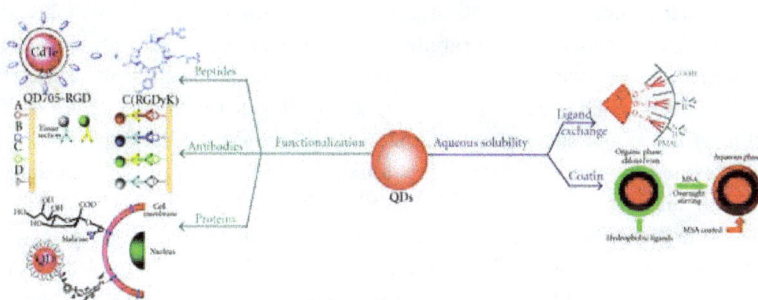

Figure 8: Strategies to deal with QDs aqueous solubility problems and fabricate functionalized QDs for biological applications.

8. QDs in biological imaging

The near infer red (NIR) emitting QDs with wavelength ranging from 700 to 900 nm are particularly interesting for biological imaging because of two reasons: 1) near-IR wavelength: has its maximum depth of penetration in tissue and 2) the interference of tissue auto-fluorescence (emission between 400 nm and 600 nm) is minimal [99]. More interestingly QDs conjugated with bio-molecules is more useful in terms of collecting specific cellular responses and well as the bio-molecule can be used to recognize some specific targets. This surface modification can also be useful to overcome the aggregation problem, minimize the nonspecific binding, and are critical to achieving specific target imaging in biological studies. Here we will discuss the recent development of conjugated QDs for in vitro and in vivo imaging.

9. *In-vitro* imaging

The initial studies of using QDs for in-vitro studies, investigators have used PbS and PbSe capped with carboxylic groups (water soluble) to label cells [100], furthermore, CdTe capped with 3-mercaptopropionic acid (3-MPA) was used as imaging tool to label

Salmonella typhimurium cells, [101] and use acid-capped CdSe/ZnS QDs for imaging HeLa cells [102]. The QDs can be stored in vesicles through endocytosis and cell tracking by using avidin-conjugated QDs to label cells. One of such works demonstrated by Gac et al. have successfully detected apoptotic cells by conjugating QDs with biotinylated Annexin V, which enables QDs to bind the phosphatidylserine (PS) moieties present on the membrane of apoptotic cells but not on other cells (Fig. 6) [103]. The silica-coating on CdTe QDs with functionalized groups is used for labeling proteins, at the same time the silica shell prevents toxic Cd^{2+} leaking from the core [104]. QDs coated with polyethylene glycol (PEG) grafted polyethylenimine (PEI), are capable of penetrating cell membranes and disrupting endosomal organelles in cells [105]. Core/shell CdSe/ZnS QDs with PEG coating and conjugated with peptide are used for cytosol localization and nucleus targeting [106,107]. Furthermore, DNA passivated CdS QDs used as new biological imaging agent [108], L-cysteine capped CdSe QDs used to label serum albumin (BSA) and cells [109], polymer QDs (PS-PEG-COOH) with conjugation to biomolecules, such as streptavidin and immunoglobulin G (IgG) to label cell surface receptors and sub-cellular structures in fixed cells [110], and near-IR InAs for imaging of specific cellular proteins [111]. Multifunctional magnetic fluorescent nanocomposite consist of silica-coated Fe_3O_4 and TGA-capped CdTe QDs has been used to successfully label and image HeLa cell line and at same time its helps magnetic separation [112]. The challenges encountered in cancer therapy stimulate the focus on the tracking and diagnosis of cancer cells. Near-IR QDs (emission wavelength of 800 nm) has been used to label squamous carcinoma cell line U14 and the fluorescent images of the cell can be clearly obtained after 6 hours by cell endocytosis [113]. A more complex QDs-aptamer- (Apt-) doxorubicin (Dox) conjugate system shows the capability of targeting, imaging, therapy, and sensing the prostate cancer cells that express the prostate-specific membrane antigen protein (Fig. 9) [114]. There are reports in literature for imaging pancreatic cancer cells by using CdSe/CdS/ZnS QDs conjugated with anti-Claudin-4 as a targeting ligand. InP QDs conjugated with cancer antibodies, and water-soluble Si-QDs micelles encapsulated with polymer chains [115]. Another important application of QDs in biological application is transfection, which is the process of deliberately introducing naked or purified nucleic acids into eukaryotic cells as reported in literature on fluorescence imaging and nucleus targeting of living cells by transfection and RNA delivery. More interesting research showed insect neuropeptide, also-called allatostatin, which can transfect with high efficiency into living human epidermoid carcinoma cells and transports quantum dots (QDs) inside the cytoplasm and even within the nucleus of the cells and thus has promise for DNA gene delivery and cell labeling, it has also been used in photodynamic therapy [116].

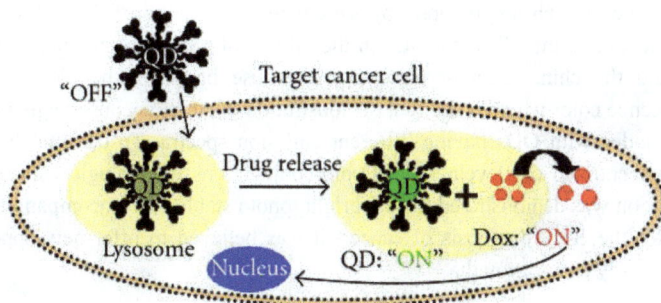

Figure 9: Schematic illustration of specific uptake of QDApt(Dox) conjugates into target cancer cell through PSMA mediate endocytosis. The release of Dox from the QD-Apt(Dox) conjugates induces the recovery of fluorescence from both QD and Dox ("ON" state), thereby sensing the intracellular delivery of Dox and enabling the synchronous fluorescent localization and killing of cancer cell (with permission from [114]).

10. *In-vivo* imaging

Apart from usage of QDs as targeting agent and labels for *in-vitro* imaging, they are also widely used as *in-vivo* imaging agents (Fig. 10) [117]. NIR QDs coated with polymer and coupled with cancer-specific antibodies is the most popular QD agent for tumor-targeted imaging. A detailed progress in this area of research has been reviewed [118]. The use of QDs *in-vivo* imaging is challenging as they must be less toxic, and should posses high contrast, high sensitivity, and good photo-stability. Phospholipid block-copolymer encapsulated CdSe/ZnS QDs (QD micelles) with functionalized DNA molecules was used to obtain Xenopus embryo fluorescent images in PBS which showed high stability and nontoxicity ($<5 \times 10^9$ nanocrystals per cell) after injection into embryos [119]. The CdSe/ZnS QDs also showed high contrast and imaging depth in two-photon excitation confocal microscopy by visualizing blood vessels in live mice [120]. The QDs coated with PEG ligand could further reduce their toxicity and accumulation in liver, thus polymer coated QDs are very useful for *in-vivo* imaging [121]. In view of toxicity of the QDs during this *in-vivo* use, non-Cd-containing QDs such as $CuInS_2$/ZnS core/shell nanocrystals have been developed and used for *in-vivo* imaging [118]. With certain concentration of QDs, fluorescence signals accumulated in the tumor could be detected after incubation of 16 days and it has been concluded that it could be compared with Computer assisted Tomography (CT) and Magnetic Resonance Imaging (MRI). QD800-based imaging could efficiently increase the sensitivity of early diagnosis of cancer cells.

Fluorescence lymph angiography by simultaneous injecting of five QDs with different emission spectra into different sites in the middle of phalanges, the upper extremity, the ears, and the chin, different parts of the mouse body can be identified by certain fluorescence color which is the first demonstration of simultaneous imaging of trafficking lymph nodes with QDs having different emission spectra. To date much attention has been concentrated on developing QD bio-conjugations for biological imaging, and this conjugation was demonstrated to be a bright, photo stable and biocompatible luminescent probe for the early diagnosis of cancer, it was believed to offer new opportunities for imaging early tumor growth.

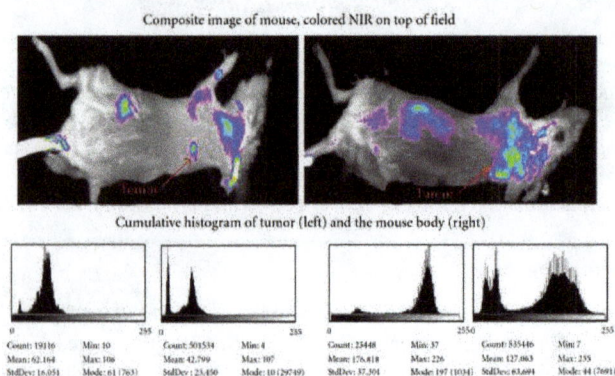

Figure 10: Targeting of lung melanoma tumor by injecting 2C5 QD-Mic in two mice.(with permission from [118]).

11. Toxicity

To consider these nanoparticles to use them in biological studies, safety, and biocompatibility are the issues that must be discussed and measured action must be taken into consideration. As suggested by toxicological studies, certain potential health effects occurred with the exposure of nanoparticles of heavy toxic metals but the quantitative studies on this matter are relatively obscure [122]. A detailed understanding is needed to better understand the effect of these nanoparticles in human cell biology. Furthermore, the toxicity of the QDs depends on factors like physicochemical condition of the cell and their environments [122-124]. To get proper idea about toxicity, several parameters of QDs need to be carefully considered such as sizes, charges, concentrations, outer coating materials and functional groups, oxidation, and mechanical stability [123] etc. Derfus et al. reported slow leakage of Cd^{2+} ions from the bare CdSe QDs due to deterioration of

QDs lattice under experimental conditions. To overcome this problem of toxicity of leaking of heavy metal from QD, surface coating and core/shell structure are considered to be efficient solutions to this problem. A secondary coating of less toxic or non-toxic material could be a choice to overcome this problem, one of this example is use of core/shell (CdSe/ZnS) and core/shell/shell (CdSe/ZnS/Silica coating) QDs [125]. The size of the particles also play an important role as nanoparticles within certain diameters of similar size of certain cellular components and proteins, which enable them to bypass natural mechanical barriers, causing tissue reactions [126].

12. Conclusion

In summary, this chapter describes a definition of colloidal nanocrystals, and their size dependent chemical and physical properties. Semiconductor colloidal nanocrystals are discussed in detail. The different synthesis methods of such materials have been highlighted. A number of properties of colloidal nanocrystals were presented, with special focus placed on their optical properties. The size dependent tunable optical properties of the semiconductor nanocrystals arising from quantum confinement promises huge potential in practical applications of these particles ranging from solar cells, photodetectors, catalysts, light emitting devices, lasers and in biological markers. In this chapter we mostly focused on QDs application in biology. The chemical and physical properties of QDs collectively suggest that they will continue to generate very active research interest from both the standpoints of fundamental science and impactful applications in biology.

8. References

[1]   S. Link, M.A. El-Sayed, Spectral Properties and Relaxation Dynamics of Surface Plasmon Electronic Oscillations in Gold and Silver Nanodots and Nanorods, J. Phys. Chem. B 40 (1999) 8410-8426. https://doi.org/10.1021/jp9917648

[2]   A.L. Efros, Interband Absorption of Light in a Semiconductor Sphere., SoV. Phys. Semicond. 16(1982) 772-775.

[3]   L. Brus, Electronic wave functions in semiconductor clusters: experiment and theory., J. Phys. Chem. 90 (1986) 2555-2560. https://doi.org/10.1021/j100403a003

[4]   J.M. Bruchez, M. Moronne, P. Gin, S.Weiss, A.P. Alivisatos, Semiconductor Nanocrystals as Fluorescent Biological Labels, Science 281(1998) 2013-2016. https://doi.org/10.1126/science.281.5385.2013

[5]   V.I. Klimov, A.A.Mikhailovsky, S. Xu, A. Malko, et al.,Optical gain and stimulated emission in nanocrystal quantum dots., Science 290 (2000) 314-317. https://doi.org/10.1126/science.290.5490.314

[6]   V.I. Klimov, Mechanisms for Photogeneration and Recombination of Multiexcitons in Semiconductor Nanocrystals: Implications for Lasing and Solar Energy Conversion. J. Phys. Chem. B. 110(2006) 16827-16845. https://doi.org/10.1021/jp0615959

[7]   V.I. Klimov, S.A. Ivanov, J. Nanda, et al. Single-exciton optical gain in semiconductor nanocrystals., Nature 447 (2007) 441-446. https://doi.org/10.1038/nature05839

[8]   S. Coe, W. Woo, M.G. Bawendi, V. Bulovic, Electroluminescence from single monolayers of nanocrystals in molecular organic devices., Nature 420(2002), 800-803. https://doi.org/10.1038/nature01217

[9]   I. Gur, N.A. Fromer, M.L. Geier, A.P. Alivisatos, Air-Stable All-Inorganic Nanocrystal Solar Cells Processed from Solution, Science 310 (2005), 462-465. https://doi.org/10.1126/science.1117908

[10]   P.K. Jain, X. Huang, I.H. El-Sayed, M.A. El-Sayad, Review of Some Interesting Surface Plasmon Resonance-enhanced Properties of Noble Metal Nanoparticles and Their Applications to Biosystems., Plasmonics 2 (2007) 107-118. https://doi.org/10.1007/s11468-007-9031-1

[11]   A.P. Alivisatos,; W.W. Gu, C. Larabell, Quantum dots as cellular probes, Annu. ReV. Biomed. Eng. 7(2005) 55-76. https://doi.org/10.1146/annurev.bioeng.7.060804.100432

[12]   R. Hergt, S. Dutz, R. Muller, M. Zeisberger, Magnetic particle hyperthermia: nanoparticle magnetism and materials development for cancer therapy. J. Phys.: Condens. Matter 18(2006) S2919-S2934. https://doi.org/10.1088/0953-8984/18/38/S26

[13]   Y. Jun, W. Seo, J. W. Cheon, A. Nanoscaling Laws of Magnetic Nanoparticles and Their Applicabilities in Biomedical Sciences. Acc. Chem. Res. 41(2008), 179-189. https://doi.org/10.1021/ar700121f

[14]   S. Laurent, D. Forge, M. Port, Magnetic Iron Oxide Nanoparticles: Synthesis, Stabilization, Vectorization, Physicochemical Characterizations, and Biological Applications., Chem. ReV. 108(2008) 2064-2110. https://doi.org/10.1021/cr068445e

[15] N. Moumen, M.P. Pileni, New Syntheses of Cobalt Ferrite Particles in the Range 2−5 nm: Comparison of the Magnetic Properties of the Nanosized Particles in Dispersed Fluid or in Powder Form., Chem. Mater. 8(1996) 1128-1134. https://doi.org/10.1021/cm950556z

[16] A.T. Ngo, M.P. Pileni, Nanoparticles of Cobalt Ferrite: Influence of the Applied Field on the Organization of the Nanocrystals on a Substrate and on Their Magnetic Properties. Adv. Mater. 12 (2000), 276-279. https://doi.org/10.1002/(SICI)1521-4095(200002)12:4<276::AID-ADMA276>3.0.CO;2-D

[17] P.A. Taleb, M.P. Pileni, Self-Organization of Magnetic Nanosized Cobalt Particles. Adv. Mater. 10(1998) 259-261. https://doi.org/10.1002/(SICI)1521-4095(199802)10:3<259::AID-ADMA259>3.0.CO;2-R

[18] T.J. Trentler, T.E. Denler, J.F. Bertone, et.al. Synthesis of TiO2 Nanocrystals by Nonhydrolytic Solution-Based Reactions., J. Am. Chem. Soc. 121 (1999) 1613-1614. https://doi.org/10.1021/ja983361b

[19] J. Joo, T. Yu, Y.W. Kim, et al. Multigram Scale Synthesis and Characterization of Monodisperse Tetragonal Zirconia Nanocrystals. J. Am. Chem. Soc. 125(2003) 6553-6557. https://doi.org/10.1021/ja034258b

[20] C.B. Murray, D.J. Norris, M.G. Bawendi, Synthesis and characterization of nearly monodisperse CdE (E = sulfur, selenium, tellurium) semiconductor nanocrystallites., J. Am. Chem. Soc. 115 (1993) 8706-8715. https://doi.org/10.1021/ja00072a025

[21] Y. Yin, A.P. Alivisatos, Colloidal nanocrystal synthesis and the organic–inorganic interface., Nature. 437(2005) 664-670. https://doi.org/10.1038/nature04165

[22] D.V. Talapin, A.L. Rogach, A. Kornowski, et al, Highly Luminescent Monodisperse CdSe and CdSe/ZnS Nanocrystals Synthesized in a Hexadecylamine−Trioctylphosphine Oxide−Trioctylphospine Mixture., Nano Lett. 1 (2001) 207-211. https://doi.org/10.1021/nl0155126

[23] B. Blackman, D.M. Battaglia, T.D.Mishima, et al., Control of the Morphology of Complex Semiconductor Nanocrystals with a Type II Heterojunction, Dots vs Peanuts, by Thermal Cycling., Chem. Mater.19 (2007) 3815-3821. https://doi.org/10.1021/cm0704682

[24]   Z.A. Peng, X. Peng, Nearly Monodisperse and Shape-Controlled CdSe
        Nanocrystals via Alternative Routes: Nucleation and Growth. J. Am. Chem. Soc.
        124(2002) 3343-3353. https://doi.org/10.1021/ja0173167

[25]   M.L. Steigerwald, A.P. Livisatos, J.M.Gibson, et al. Surface derivatization and
        isolation of semiconductor cluster molecules. J. Am. Chem. Soc. 110 (1988) 3046-
        3050. https://doi.org/10.1021/ja00218a008

[26]   G. Morello, Giorgi M. De, S, Kudera, et al. Temperature and Size Dependence of
        Nonradiative Relaxation and Exciton–Phonon Coupling in Colloidal CdTe
        Quantum Dots. J. Phys. Chem. C. 111 (2007) 5846-5849.
        https://doi.org/10.1021/jp068307t

[27]   K.-T. Yong, Y. Sahoo, M.T. Swihart, et al. Shape Control of CdS Nanocrystals in
        One-Pot Synthesis. J. Phys. Chem. C 111 (2007) 2447-2458.
        https://doi.org/10.1021/jp066392z

[28]   O.I. Micic, C.J. Curtis, K.M. Jones, et al., Synthesis and Characterization of InP
        Quantum Dots. J. Phys. Chem. 98(1994) 4966-4969.
        https://doi.org/10.1021/j100070a004

[29]   S.P. Ahrenkiel, O.I. Mićić, A. Miedaner, C.J. Curtis, J.M. Nedeljković, and A.J.
        Nozik. Synthesis and Characterization of Colloidal InP Quantum Rods., Nano
        Letters, 3(2003) 833–837. https://doi.org/10.1021/nl034152e

[30]   D.V. Talapin, A.L. Rogach, E.V.Shevchenko, et al. Dynamic Distribution of
        Growth Rates within the Ensembles of Colloidal II–VI and III–V Semiconductor
        Nanocrystals as a Factor Governing Their Photoluminescence Efficiency., J. Am.
        Chem. Soc. 124(2002) 5782-5790. https://doi.org/10.1021/ja0123599

[31]   D. Battaglia, X. Peng, Formation of High Quality InP and InAs Nanocrystals in a
        Noncoordinating Solvent., Nano Lett. 2(2002) 1027-1030.
        https://doi.org/10.1021/nl025687v

[32]   M.A. Hines, G.D. Scholes, Colloidal PbS Nanocrystals with Size-Tunable Near-
        Infrared Emission: Observation of Post-Synthesis Self-Narrowing of the Particle
        Size Distribution., Advanced Materials. 15 (2003) 1844-1849.
        https://doi.org/10.1002/adma.200305395

[33]   D.V. Talapin, C.B. Murray, PbSe Nanocrystal Solids for n- and p-Channel Thin
        Film Field-Effect Transistors., Science 310(2005) 86-89.
        https://doi.org/10.1126/science.1116703

[34] J.J. Urban, D.V.Talapin, E.V.Shevchenko, et al. Self-Assembly of PbTe Quantum Dots into Nanocrystal Superlattices and Glassy Films, J. Am.Chem. Soc. 128 (2006) 3248-3255. https://doi.org/10.1021/ja058269b

[35] D.V. Talapin, J.S. Lee, M.V. Kovalenko, E.V. Shevchenko, Prospects of Colloidal Nanocrystals for Electronic and Optoelectronic Applications., Chem. Rev. 110 (2010) 389–458. https://doi.org/10.1021/cr900137k

[36] M.A. Hines, P. Guyot-Sionnest, Synthesis and Characterization of Strongly Luminescing ZnS-Capped CdSe Nanocrystals., J. Phys Chem, 100 (1996) 468-471. https://doi.org/10.1021/jp9530562

[37] B. Andrew Greytak, M. Allen Peter, Liu Wenhao, et al., Alternating layer addition approach to CdSe/CdS core/shell quantum dots with near-unity quantum yield and high on-time fractions., Chem. Sci., 3(2012) 2028–2034. https://doi.org/10.1039/c2sc00561a

[38] N.R. Jana, X. Peng, Single-Phase and Gram-Scale Routes toward Nearly Monodisperse Au and Other Noble Metal Nanocrystals., J. Am. Chem. Soc. 125 (2003) 14280-14281. https://doi.org/10.1021/ja038219b

[39] M. Brust, M. Walker, D. Bethell, et al., Synthesis of thiol-derivatised gold nanoparticles in a two-phase Liquid–Liquid system., Chem. Commun. 7(1994) 801-802. https://doi.org/10.1039/C39940000801

[40] S.E. Skrabalak, B.J. Wiley, M.H. Kim, et al., On the Polyol Synthesis of Silver Nanostructures: Glycolaldehyde as a Reducing Agent., Nano Lett. 8(2008) 2077-2081. https://doi.org/10.1021/nl800910d

[41] B. Wiley, S.-H. Im, Z.-Y Li, J.M. McLellan, A. Siekkinen, Y.J. Xia, Maneuvering the surface plasmon resonance of silver nanostructures through shape-controlled synthesis., Phys. Chem. B 110(2006) 15666-15675. https://doi.org/10.1021/jp0608628

[42] Y. Borodko, S.M. Humphrey, T. Don Tilley, H. Frei, G.A. Somorjai, Charge-Transfer Interaction of Poly(vinylpyrrolidone) with Platinum and Rhodium Nanoparticles., J. Phys. Chem. C 111(2007) 6288-6295. https://doi.org/10.1021/jp068742n

[43] C. Wang, H. Daimon, T. Onodera, T. Koda, S. Sun, A general approach to the size- and shape-controlled synthesis of platinum nanoparticles and their catalytic reduction of oxygen., Angew. Chem., Int. Ed. 47(2008) 3588-3591. https://doi.org/10.1002/anie.200800073

[44] S.U. Son, Y. Jang, K.Y.Yoon, E. Kang, T. Facile Hyeon, Synthesis of Various Phosphine-Stabilized Monodisperse Palladium Nanoparticles through the Understanding of Coordination Chemistry of the Nanoparticles., Nano Lett. 4(2004), 1147-1151. https://doi.org/10.1021/nl049519+

[45] Y. Sun, B. Wiley, Z.-Y. Li, Y. Xia, Synthesis and Optical Properties of Nanorattles and Multiple-Walled Nanoshells/Nanotubes Made of Metal Alloys., J. Am. Chem. Soc. 126(2004) 9399-9406. https://doi.org/10.1021/ja048789r

[46] Y. Yin, A.P. Alivisatos, Colloidal nanocrystal synthesis and the organic–inorganic interface., Nature 437(2005) 664-670. https://doi.org/10.1038/nature04165

[47] C.D.M. Donega, Synthesis and properties of colloidal heteronanocrystals., Chem. Soc. Rev., 40(2011) 1512–1546. https://doi.org/10.1039/C0CS00055H

[48] Y. Wang, N. Herron, Nanometer-sized semiconductor clusters: materials synthesis, quantum size effects, and photophysical properties., J. Phys. Chem. 95(1991) 525–532. https://doi.org/10.1021/j100155a009

[49] J. Bang, H. Yang, P.H. Holloway, Enhanced and stable green emission of ZnO nanoparticles by surface segregation of Mg., Nanotechnology 17(2006) 973–978. https://doi.org/10.1088/0957-4484/17/4/022

[50] E. Kucur, W. Bucking, R. Giernoth, et al., Determination of Defect States in Semiconductor Nanocrystals by Cyclic Voltammetry., Phys. Chem. B 109(2005) 20355–20360. https://doi.org/10.1021/jp053891b

[51] A.M. Smith, S. Nie, Semiconductor Nanocrystals: Structure, Properties, and Band Gap Engineering., Acc. Chem. Res.43(2010) 190–200. https://doi.org/10.1021/ar9001069

[52] U.E.H. Laheld, F.B. Pedersen, P.C. Hemmer, Excitons in type-II quantum dots: Finite offsets, Phys. Rev. B 52(1995) 2697-2703. https://doi.org/10.1103/PhysRevB.52.2697

[53] M.V. Kovalenko, M. Scheele, D.V0. Talapin, Colloidal nanocrystals with molecular metal chalcogenide surface ligands, Science 324 (2009) 1417-1420. https://doi.org/10.1126/science.1170524

[54] A.L. Efros, M. Random Rosen, Telegraph Signal in the Photoluminescence Intensity of a Single Quantum Dot., Phys. Rev. Lett. 78 (1997) 1110–1113. https://doi.org/10.1103/PhysRevLett.78.1110

[55] E.W. Williams, R. Hall, Pergomon Press: New York, NY, USA, 1977; p 237.

[56] M. Nirmal, B.O. Dabbousi, M.G. Bawendi, et al., Fluorescence intermittency in single cadmium selenide nanocrystals., Nature 383(1996) 802–804. https://doi.org/10.1038/383802a0

[57] Y. Chen, J. Vela, H. Htoon, et al., "Giant" multishell CdSe nanocrystal quantum dots with suppressed blinking., J. Am. Chem. Soc. 130(2008) 5026-5027. https://doi.org/10.1021/ja711379k

[58] C. Galland, Y. Ghosh, A. Steinbrück, M. Sykora, et al., Two types of luminescence blinking revealed by spectroelectrochemistry of single quantum dots., Nature 479(2011) 203-207. https://doi.org/10.1038/nature10569

[59] A.L. Efros, M. Rosen, Random Telegraph Signal in the Photoluminescence Intensity of a Single Quantum Dot., Phys. Rev. Lett. 78 (1997) 1110 -1113. https://doi.org/10.1103/PhysRevLett.78.1110

[60] C. Galland, Y. Ghosh, A. Steinbrück, J.A. Hollingsworth, et al., Lifetime blinking in nonblinking nanocrystal quantum dots., Nature Commun. 3 (2012) 908 (7pages).

[61] A.K. Bhattacharjee, Optical properties of paramagnetic ion-doped semiconductor nanocrystals., Phys. Rev. B: Condens. Matter Mater.Phys., 68(2003) 045303 (6 pages).

[62] S.L. Cumberland, K.M. Hanif, A. Javier, et al., Inorganic Clusters as Single-Source Precursors for Preparation of CdSe, ZnSe, and CdSe/ZnS Nanomaterials., Chem. Mater.14(2002) 1576–1584. https://doi.org/10.1021/cm010709k

[63] S.C. Erwin, L.J. Zu, M.I. Ha, et al., Doping semiconductor nanocrystals., Nature 436 (2005) 91–94. https://doi.org/10.1038/nature03832

[64] R.N. Bhargava, D. Gallagher, X. Hong, A. Nurmikko, Optical properties of manganese-doped nanocrystals of ZnS., Phys. Rev.Lett. 72(1994) 416–419. https://doi.org/10.1103/PhysRevLett.72.416

[65] A. Nag, S. Sapra, C. Nagamani, et al., Study of Mn2+ Doping in CdS Nanocrystals., Chem. Mater., 19(2007) 3252–3259. https://doi.org/10.1021/cm0702767

[66] J.W. Stouwdam, R.A.J. Janssen, Electroluminescent Cu-doped CdS Quantum Dots., Adv. Mater., 21 (2009) 2916–2920. https://doi.org/10.1002/adma.200803223

[67] F. Zhang, X.W. He, W.Y. Li, Y.K. Zhang, One-pot aqueous synthesis of composition-tunable near-infrared emitting Cu-doped CdS quantum dots as

fluorescence imaging probes in living cells., J. Mater.Chem., 22 (2012) 22250–22257. https://doi.org/10.1039/c2jm33560c

[68]   F.C. Mikulec, M. Kuno, M. Bennati, D.A. Hall, R.G. Griffin, G.M. Bawendi, Organometallic Synthesis and Spectroscopic Characterization of Manganese-Doped CdSe Nanocrystals. J. Am. Chem. Soc., 122 (2000) 2532–2540. https://doi.org/10.1021/ja991249n

[69]   B.B. Srivastava, S. Jana, N. Pradhan, Doping Cu in Semiconductor Nanocrystals: Some Old and Some New Physical Insights., J. Am. Chem. Soc., 133(2011) 1007–1015. https://doi.org/10.1021/ja1089809

[70]   S. Jana, B.B. Srivastava, R. Bose, N. Pradhan,. Multifunctional Doped Semiconductor Nanocrystals. J. Phys. Chem. Lett., 3 (2012) 2535–2540. https://doi.org/10.1021/jz3010877

[71]   R.G. Xie, X.G. Peng, Synthesis of Cu-Doped InP Nanocrystals (d-dots) with ZnSe Diffusion Barrier as Efficient and Color-Tunable NIR Emitters., J. Am. Chem. Soc., 131(2009)10645–10651. https://doi.org/10.1021/ja903558r

[72]   J.R. Lakowicz, I. Gryczynski, Z. Gryczynski, C.J. Murphy, Luminescence Spectral Properties of CdS Nanoparticles., J. Phys. Chem. B, 103 (1999) 7613–7620. https://doi.org/10.1021/jp991469n

[73]   T. Torimoto, M. Yamashita, S. Kuwabata, et al., Fabrication of CdS Nanoparticle Chains along DNA Double Strands., J. Phys. Chem. B 103(1999) 8799–8803. https://doi.org/10.1021/jp991781x

[74]   J.R. Lakowicz, I. Gryczynski, Z. Gryczynski, et al., Temperature- and salt-dependent binding of long DNA to protein-sized quantum dots: thermodynamics of "inorganic protein"-DNA interactions., Anal. Biochem. 280(2000) 128–136. https://doi.org/10.1006/abio.2000.4495

[75]   S.R. Bigham, J.L. Coffer, Thermochemical Passivation of DNA-Stabilized Q-Cadmium Sulfide Nanoparticles., J. Cluster Sci. 11 (2000) 359–372. https://doi.org/10.1023/A:1009049823345

[76]   X. Li, J.L.Coffer, Effect of Pressure on the Photoluminescence of Polynucleotide-Stabilized Cadmium Sulfide Nanocrystals., Chem. Mater, 11(1999) 2326–30. https://doi.org/10.1021/cm980485e

[77]   I. Willner, F. Patolsky, J. Wasserman, Nanoparticles, Proteins, and Nucleic Acids: Biotechnology Meets Materials Science., Angew. Chem. Int. Ed. Engl. 40(2000)

1861–1864. https://doi.org/10.1002/1521-3773(20010518)40:10<1861::AID-ANIE1861>3.0.CO;2-V

[78] I. Sondi, O. Siiman, S. Koester, E. Matijevic, Preparation of Aminodextran−CdS Nanoparticle Complexes and Biologically Active Antibody−Aminodextran−CdS Nanoparticle Conjugates., Langmuir 16(2000) 3107–3118. https://doi.org/10.1021/la991109r

[79] H.M. Chen, X.F. Huang, L. Xu, J. Xu, K.J. Chen, D. Feng, Self-assembly and photoluminescence of CdS-mercaptoacetic clusters with internal structures., Superlatt.Microstruct. 27 (2000) 1–5. https://doi.org/10.1006/spmi.1999.0794

[80] A.L. Rogach, L. Katsikas, A. Kornowski, D. Su, A. Eychmueller, H. Weller, Synthesis and Characterization of a Size Series of Extremely Small Thiol-Stabilized CdSe Nanocrystals., J. Phys. Chem. B, 103(1999) 3065–3069. https://doi.org/10.1021/jp984833b

[81] N.N. Mamedova, N.A. Kotov, Albumin−CdTe Nanoparticle Bioconjugates: Preparation, Structure, and Interunit Energy Transfer with Antenna Effect., Nano Lett. 1(2001) 281–286. https://doi.org/10.1021/nl015519n

[82] F. Pinaud, D. King, H.P. Moore, and S. Weiss, Bioactivation and cell targeting of semiconductor CdSe/ZnS nanocrystals with phytochelatin-related peptides., J. Am. Chem. Soc. 126 (2004), , 6115–6123. https://doi.org/10.1021/ja031691c

[83] W. Chan and M. Nie, Quantum dot bioconjugates for ultrasensitive nonisotopic detection., Science 281(1998) 2016–2018. https://doi.org/10.1126/science.281.5385.2016

[84] T. Pellegrino, L. Manna, S.T. Kudera Liedl., et al., Hydrophobic Nanocrystals Coated with an Amphiphilic Polymer Shell: A General Route to Water Soluble Nanocrystals., Nano Letters 4(2004) 703–707. https://doi.org/10.1021/nl035172j

[85] C.A.J. Lin, R.A. Sperling, J.K. Li, et al., Design of an Amphiphilic Polymer for Nanoparticle Coating and Functionalization., Small 4 (2008) 334–341. https://doi.org/10.1002/smll.200700654

[86] G.P. Mitchell, C.A. Mirkin and R.L. Letsinger, Programmed Assembly of DNA Functionalized Quantum Dots., J. Am. Chem. Soc. 121(1999) 8122–8123. https://doi.org/10.1021/ja991662v

[87] C-.C. Chen, C-.P. Yet, H-.N. Wang and C-.Y.Chao, Water-Soluble, Isolable Gold Clusters Protected by Tiopronin and Coenzyme A Monolayers., Langmuir 15(1999) 6845–50.

[88] C.Y. Zhang, H. Ma, S.M. Nie, Y. Ding, L. Jin and D.Y.Chen, Quantum dot-labeled trichosanthin., Analyst 125(2000) 1029–1031.
https://doi.org/10.1039/b002666m

[89] B. Sun, W. Xie, G. Yi, D. Chen, Y. Zhou and J. Cheng, Microminiaturized immunoassays using quantum dots as fluorescent label by laser confocal scanning fluorescence detection., J. Immunol. Methods 249(2001) 85–89.
https://doi.org/10.1016/S0022-1759(00)00331-8

[90] D.M. Willard, L.L. Carillo, J. Jung and A.V. Orden, CdSe−ZnS Quantum Dots as Resonance Energy Transfer Donors in a Model Protein−Protein Binding Assay, Nano Lett. 1(2001) 469–474. https://doi.org/10.1021/nl015565n

[91] J. Aldana, Y.A. Wang and X. Peng, Photochemical Instability of CdSe Nanocrystals Coated by Hydrophilic Thiols., J. Am. Chem. Soc. 123(2001) 8844–8850. https://doi.org/10.1021/ja016424q

[92] D. Gerion, F. Pinaud, S.C. Williams, W.J. Parak, D. Zanchet, S Weiss and A.P. Alivisatos, Synthesis and Properties of Biocompatible Water-Soluble Silica-Coated CdSe/ZnS Semiconductor Quantum Dots., J. Phys. Chem. B. 105(2001) 8861–8871. https://doi.org/10.1021/jp0105488

[93] H. Mattoussi, J.M. Mauro, E.R. Goldman, G.P. Anderson, V.C. Sundar, F.V. Mikulec and M.G. Bawendi, Self-Assembly of CdSe−ZnS Quantum Dot Bioconjugates Using an Engineered Recombinant Protein., J. Am.Chem. Soc. 122 (2000) 12142–12150. https://doi.org/10.1021/ja002535y

[94] L. Guo, S. Yang, C. Yang, et al., Highly monodisperse polymer-capped ZnO nanoparticles: Preparation and optical properties., Appl. Phys. Lett. 76(2000) 2901–2903. https://doi.org/10.1063/1.126511

[95] L.M. Liz-Marz´an, M. Giersig and P. Mulvaney, Synthesis of Nanosized Gold−Silica Core−Shell Particles., Langmuir. 12 (1996) 4329–4335.
https://doi.org/10.1021/la9601871

[96] P. Mulvaney, L.M. Liz-Marz´an, M. Giersig and T. Ung, Silica encapsulation of quantum dots and metal clusters., J. Mater. Chem. 10(2000) 1259–1270.
https://doi.org/10.1039/b000136h

[97] W.J. Parak, D. Gerion, D. Zanchet, et al., Conjugation of DNA to Silanized Colloidal Semiconductor Nanocrystalline Quantum Dots. Chem. Mater. 14 (2002) 2113–2119. https://doi.org/10.1021/cm0107878

[98] A. Schroedter, H. Weller, Biofunctionalization of Silica-Coated CdTe and Gold Nanocrystals., J M. Nano Lett. 2(2002) 1363–1367. https://doi.org/10.1021/nl025779k

[99] W. Jiang, A. Singha., J.N. Zheng, et al., Optimizing the Synthesis of Red- to Near-IR-Emitting CdS-Capped CdTexSe1-x Alloyed Quantum Dots for Biomedical Imaging. Chemistry of Materials, 18 (2006) 4845–4854. https://doi.org/10.1021/cm061311x

[100] B.R. Hyun, H.Y. Chen, D.A. Rey, F.W. Wise, and C.A. Batt, Near-Infrared Fluorescence Imaging with Water-Soluble Lead Salt Quantum Dots. Journal of Physical Chemistry B, 111(2007) 5726–5730. https://doi.org/10.1021/jp068455j

[101] H. Li, W.Y. Shih, and W.H. Shih, Synthesis and Characterization of Aqueous Carboxyl-Capped CdS Quantum Dots for Bioapplications., Industrial and Engineering Chemistry Research, 46(2007) 2013–2019. https://doi.org/10.1021/ie060963s

[102] J.K. Jaiswal, H. Mattoussi, J.M. Mauro, and S.M. Simon, Long-term multiple color imaging of live cells using quantum dot bioconjugates. Nature Biotechnology, 21(2002) 47–51. https://doi.org/10.1038/nbt767

[103] S.L. Gac, I. Vermes, and A.V.D. Berg, Quantum Dots Based Probes Conjugated to Annexin V for Photostable Apoptosis Detection and Imaging., Nano Letters, 6 (2006) 1863–1869. https://doi.org/10.1021/nl060694v

[104] A. Wolcott, D. Gerion, M. Visconte, et al., Adam. Silica-Coated CdTe Quantum Dots Functionalized with Thiols for Bioconjugation to IgG Proteins. Journal of Physical Chemistry B, 110 (2006) 5779–5789. https://doi.org/10.1021/jp057435z

[105] H.W. Duan and S.M. Nie, Cell-Penetrating Quantum Dots Based on Multivalent and Endosome-Disrupting Surface Coatings., J. Am. Chem. Soc., 129 (2007) 3333–3336. https://doi.org/10.1021/ja068158s

[106] F.Q. Chen and D. Gerion, Nano Letters, Fluorescent CdSe/ZnS Nanocrystal−Peptide Conjugates for Long-term, Nontoxic Imaging and Nuclear Targeting in Living Cells. 4(2004)1827–1832.

[107] I. Yildiz, B. McCaughan, S.F. Cruickshank, J.F. Callan, and F.M. Raymo, Biocompatible CdSe−ZnS Core−Shell Quantum Dots Coated with Hydrophilic Polythiols., Langmuir 25 (2009)7090–7096. https://doi.org/10.1021/la900148m

[108] N. Ma, J. Yang, K.M. Stewart, and S.O. Kelley, DNA-Passivated CdS Nanocrystals: Luminescence, Bioimaging, and Toxicity Profiles. Langmuir, 23 (2007) 12783–12787. https://doi.org/10.1021/la7017727

[109] P. Liu, Q.S. Wang, and X. Li, Studies on CdSe/l-cysteine Quantum Dots Synthesized in Aqueous Solution for Biological Labeling. Journal of Physical Chemistry C, 113(2009) 7670–7676. https://doi.org/10.1021/jp901292q

[110] C.F. Wu, T. Schneider, M. Zeigler, et al., Bioconjugation of Ultrabright Semiconducting Polymer Dots for Specific Cellular Targeting., J. Am. Chem. Soc. 132(2010) 15410–15417. https://doi.org/10.1021/ja107196s

[111] P.M. Allen, W.H. Liu, V.P. Chauhan, et al., InAs(ZnCdS) Quantum Dots Optimized for Biological Imaging in the Near-Infrared., J. Am. Chem. Soc. 132 (2010) 470–471. https://doi.org/10.1021/ja908250r

[112] P. Sun, H.Y. Zhang, C. Liu, et al., Langmuir 26(2010) 1278–1284. https://doi.org/10.1021/la9024553

[113] Y.A. Cao, K. Yang, Z.G. Li, et al., Near-infrared quantum-dot-based non-invasive in vivo imaging of squamous cell carcinoma U14. Nanotechnology 21(2010) Article ID 475104. https://doi.org/10.1088/0957-4484/21/47/475104

[114] V. Bagalkot, L.F. Zhang, E. Levy-Nissenbaum, et al., Quantum Dot–Aptamer Conjugates for Synchronous Cancer Imaging, Therapy, and Sensing of Drug Delivery Based on Bi-Fluorescence Resonance Energy Transfer., Nano Letters, 7 (2007) 3065–3070. https://doi.org/10.1021/nl071546n

[115] F. Erogbogbo, K.T. Yong, I. Roy, G.X. Xu, P.N. Prasad, and M.T. Swihart, Biocompatible Luminescent Silicon Quantum Dots for Imaging of Cancer Cells. ACS Nano, 2 (2008) 873–878. https://doi.org/10.1021/nn700319z

[116] V. Biju, D. Muraleedharan, K.I. Nakayama, et al., Quantum dot-Insect Neuropeptide Conjugates for Fluorescence Imaging, Transfection, and Nucleus Targeting of Living Cells., Langmuir. 23 (2007) 10254–10261. https://doi.org/10.1021/la7012705

[117] H.S. Choi, W.H. Liu, F.B. Liu, et al., Design considerations for tumour-targeted nanoparticles. Nature Nanotechnology, 5(2010) 42–47. https://doi.org/10.1038/nnano.2009.314

[118] A. Papagiannaros, J. Upponi, W. Hartner, D. Mongayt, T. Levchenko, and V. Torchilin, Quantum dot loaded immunomicelles for tumor imaging, BMCMedical Imaging 10 (2010) article 22. https://doi.org/10.1186/1471-2342-10-22

[119] B. Dubertret, P. Skourides, D.J. Norris, V. Noireaux, A.H. Brivanlou, and A. Libchaber, In vivo imaging of quantum dots encapsulated in phospholipid micelles. Science, 298(2002) 1759–1762. https://doi.org/10.1126/science.1077194

[120] B. Ballou., B.C. Lagerholm, L.A. Ernst, M.P. Bruchez, and A.S Waggoner, Noninvasive Imaging of Quantum Dots in Mice., Bioconjugate Chemistry 15(2004) 79–86. https://doi.org/10.1021/bc034153y

[121] T.J. Daou, L. Li, P. Reiss, V. Josserand, and I. Texier, Effect of Poly(ethylene glycol) Length on the in Vivo Behavior of Coated Quantum Dots., Langmuir 25 (2009) 3040–3044. https://doi.org/10.1021/la8035083

[122] A.M. Derfus, W.C.W. Chan, and S.N. Bhatia, Probing the Cytotoxicity of Semiconductor Quantum Dots., Nano Letters 4 (2004) 11–18. https://doi.org/10.1021/nl0347334

[123] R. Hardman, A Toxicologic Review of Quantum Dots: Toxicity Depends on Physicochemical and Environmental Factors., Environmental Health Perspectives 114 (2006) 165–172. https://doi.org/10.1289/ehp.8284

[124] G.N. Guo, W. Liu, J.G. Liang, et al., Preparation and characterization of novel CdSe quantum dots modified with poly (D, L-lactide) nanoparticles., Materials Letters, 60 (2006) 565–2568. https://doi.org/10.1016/j.matlet.2006.01.073

[125] J.H. Park, L. Gu, G. Von Maltzahn, E. Ruoslahti, S.N. Bhatia, and M.J. Sailor, Biodegradable luminescent porous silicon nanoparticles for in vivo applications., Nature Materials 8 (2009) 331–336. https://doi.org/10.1038/nmat2398

[126] Y. Pan, S. Neuss, A. Leifert, et al., Size-dependent cytotoxicity of gold nanoparticles., Small 3 (2007) 1941–1949. https://doi.org/10.1002/smll.200700378

Chapter 3

Synthesis of Nanoparticles through Thermal Decomposition of Organometallic Materials and Application for Biological Environment

Ashis Bhattacharjee[1] and Madhusudan Roy[2]*

[1]Department of Physics, Institute of Science, Visva-Bharati University, Santiniketan 731235, India

[2]Surface Physics and Material Science Division, Saha Institute of Nuclear Physics, 1/ AF, Bidhannagar, Kolkata, India

*madhusudan.roy@saha.ac.in

Abstract

Materials with structure at the nanoscale often offer unique optical, electronic, or mechanical properties and in turn find applications in different areas. The state of the art activities in nano-science and nano-technology have sparked the expectations to run into the challenges in the field of renewable energy, medicine and environment surveillance. There are different methods to grow nano-structures and the synthesis of nanoparticles is broadly classified under two processes such as 'Top Down' and 'Bottom Up'. Among different synthesis techniques for preparing metal/oxide nanoparticles, thermal decomposition of organometallic compounds at relatively low temperatures becomes increasingly important and useful.

Keywords

Nanoscale, Nanoparticle, Thermal Decomposition, Organometallic Compound, Metal/Oxide Nanoparticle

Contents

1. Introduction..51
   1.1 Solid state thermal decomposition..52
   1.2 $\{XR_4[M(II)M(III)(C_2O_4)_3]\}_\infty$ molecule-based materials as precursors...53
   1.3 $[(C_5H_5)_2Fe]$ materials as precursors...54

2.   **Materials** **55**

    2.1   Synthesis of $\{XR_4[M(II)M(III)(C_2O_4)_3]\}_\infty$ type precursors for thermal decomposition 55

    2.2   Synthesis of ferrocene-based precursors for thermal decomposition 58

    2.3   Synthesis of thermally decomposed material 58

    2.4   Characterization 59

3.   **Results** **59**

    3.1   Nanomaterials obtained from $\{XR_4[M(II)M(III)(C_2O_4)_3]\}_\infty$ type precursors 59

    3.1.1 $\{N(n\text{-}C_4H_9)_4[M(II)Fe(III)(C_2O_4)_3]\}_\infty$, M = Zn, Co, Fe 59

    3.1.2 $\{N(n\text{-}C_5H_{11})_4[Mn(II)Fe(III)(C_2O_4)_3]\}_\infty$ 62

    3.1.3 $\{As(C_6H_5)_4[Fe(II)Fe(III)(C_2O_4)_3]\}_\infty$ 63

    3.1.4 $\{N(n\text{-}C_4H_9)_4[Fe(II)Cr(III)(C_2O_4)_3]\}_\infty$ 64

    3.1.5 Ferrocene with Oxalic acid dihydrate 64

    3.1.6 Ferrocene with Fe(II)acetylacetonate 65

4.   **Applications for biological environment** **66**

References **67**

1.   Introduction

The human beings instinctively belong to a scale of the visual world which ranges from millimetre to meter, a span of three orders of magnitudes. If one ponders, one could find the human technology started its journey, in a sense, since the Neolithic Age and the anthropological studies bear the signatures form when humans made tools of stones for their survival and sustenance. The point of emphasis is that human got involved to develop technology in the order of his body parts. With the progress of knowledge (science), technology has expanded along the dimensions below and above the dimension of his body parts. In the present scale of technology, the scientists and the technocrats are engrossed in nano-scale activities to carry forward the torch of ever increasing development that covers not only human development, but also environment.

Nano-sized materials describe, in principle, the materials of which a single unit is sized (in at least one dimension) between 1 and 1000 nanometres (nm) but is usually 1-100 nm (according to the most accepted definition of nanoscale). Nanomaterials research leads a material-science-based approach to nanotechnology, leveraging advances in materials metrology and synthesis which have been developed in support of microfabrication research. The materials with structure at the nanoscale often have unique optical, electronic, or mechanical properties. Nanomaterials are becoming commercialized and emerging as commodities. The recent activities in nano-science and nano-technology have kindled the hopes to meet the challenges in the field of renewable energy, medicine and environment surveillance. Thus, the applications of nanomaterials are manifold.

There are various ways to develop nano-structures and a great deal of information can be gathered in the works primarily of two groups led by C.J. Murphy [1-4] and Y. Xia [5-7]. Arrays of conventional methods have been employed in the synthesis of nanoparticles. But these conventional methods are bound by various limitations such as expenses, generations of hazardous toxic chemicals, etc. To circumvent such unwarranted predicaments, major researches are directed to develop safe, eco-friendly alternative approaches in synthesis of nanoparticles among which biological systems have been focussed and exploited as a preferred green principle process for the synthesis of nanoparticles. Methods employed for the synthesis of nanoparticles are broadly classified under two processes such as a 'Top Down' process and 'Bottom Up'.

- 'Top Down' approach: Bulk material is broken down into particles at nanoscale with various lithographic techniques (e.g. grinding, milling, etching, electro-explosion, sputtering, laser ablation).

- 'Bottom Up' approach: Atoms self-assemble to new nuclei which grow into a particle of nanoscale (e.g., superficial fluid synthesis, spinning, plasma or flame spraying synthesis, sol-gel process, chemical vapour deposition, atomic or molecular condensation, green synthesis).

1.1 Solid state thermal decomposition

One can also broadly divide all the chemical synthesis processes into two classes – (i) Synthesis through wet chemistry and (ii) Synthesis through dry chemistry. In our laboratories (the authors belong to different institutions), we have adopted the thermal decomposition technique which belong to 'Synthesis through dry chemistry' (i.e., solid state thermal decomposition) to grow nano-structures of various oxides and these structures exhibit many facets of panoramic beauty.

Among different reported synthesis methods for preparing metal/oxide nanoparticles, thermal decomposition of organometallic compounds at relatively low temperatures becomes increasingly important [8-12]. Thermal decomposition of iron-bearing precursors is a very popular way to produce iron oxides of different phases [13, 14]. Synthesis of magnetic oxide nanoparticles by thermal decomposition of iron-organic compounds has several advantages, such as a relatively low temperature of formation of magnetic oxides, a short reaction time and the possibility of using inexpensive iron-organic compounds. The properties of the end decomposition product depend on the chemical nature of the precursor used, temperature and reaction atmosphere.

1.2 $\{XR_4[M(II)M(III)(C_2O_4)_3]\}_\infty$ molecule-based materials as precursors

Among many molecule-based materials, the oxalate ligand bridged homo/heterometallic networks of general molecular formula $\{XR_4[M(II)M(III)(C_2O_4)_3]\}_\infty$, where XR_4^+ = the mono positive organic cation (X = N, As; R = n-alkyl / aryl), M(II)/M(III) = di-/trivalent transition metal and $(C_2O_4)^-$ = oxalate ligand have drawn much research interest in the field of molecular magnetism, as their magnetic property range within para-, ferro-, ferri- or antiferromagnetic long-range ordering behavior [15]. The notable feature of these materials is that a wide diversity of metal ion arrangements may be suitably combined into the same basic structure made up of infinitely extended layers of oxalate-bridged metal pairs M(II) and M(III) exhibiting hexagonal symmetry and being devided by the organic cation XR_4^+. The oxalate ions serve as the bridging ligand for the transition-metal centres. A schematic representation of the material system $\{XR_4[M(II)M(III)(C_2O_4)_3]\}_\infty$ shown below (Fig. 1) provides an impression of the infinitely extended [–M(II)–oxalate–M(III)–] layer in such type of materials in which the mono positive organic cations XR_4^+ are interleaved within the interlayer space (not shown). Since the first demonstration of molecular design of structurally 2D bimetallic oxalate-bridged networks based on trioxalatochromate(III) [16] building blocks in 1992 and trioxalatoferrate building blocks [17] in 1993 by Ōkawa and his collaborators. The ease of synthesis and the tunability of the magnetic property of this system of materials led to the synthesis of a large number of magnetic materials of analogous bimetallic assemblies, also with mixed-valency stoichiometries, have been reported [15]. In 2007, Neo $et\ al.$ [18] showed that thermal decomposition of ferromagnetic material $NBu_4[Mn(II)Cr(III)(C_2O_4)_3]$ resulted in a spinel complex, $Mn_{1.5}Cr_{1.5}O_4$ at ~500°C and under suitable conditions both the metals in such heterometallic molecular complexes, if used as precursor, could be transformed into metal oxides. This prompted us to explore the solid state thermal decomposition of materials like - $\{XR_4[M(II)Fe(III)(C_2O_4)_3]\}_\infty$ [19-22], $\{N(n\text{-}C_4H_9)_4[Fe(II)_{1-x}Zn(II)_xFe(III)(C_2O_4)_3]\}_\infty$ and $\{N(n\text{-}C_4H_9)_4[Fe(II)Fe(III)_xCr(III)_{1-x}(C_2O_4)_3]\}_\infty$ [23] in order to find metal oxide nanoparticles by 'Synthesis through dry chemistry' method.

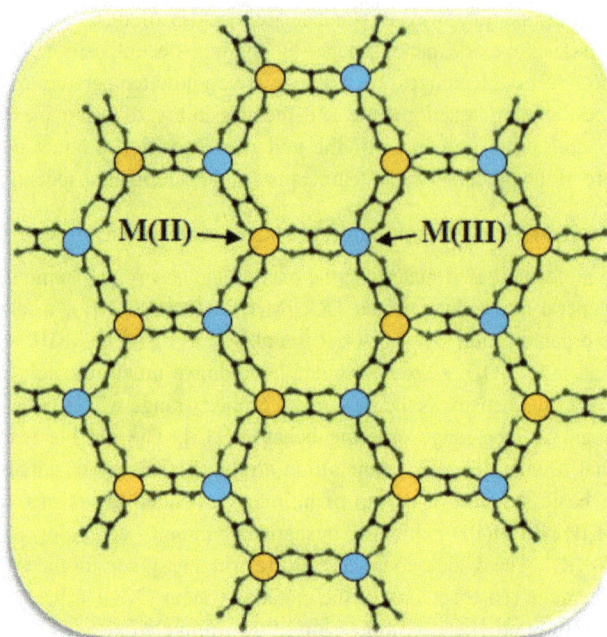

Figure 1. Schematic projection of infinitely extended –M(II)-ox-M(III)- layer of {XR$_4$[M(II)M(III)(C$_2$O$_4$)$_3$]}$_\infty$. The cations are interleaved within the interlayer spaces (not shown).

1.3   [(C$_5$H$_5$)$_2$Fe] materials as precursors

Thermal decomposition of organic iron precursors, in particular ferrocene [(C$_5$H$_5$)$_2$Fe] materials (Fig. 2), has been demonstrated to be very successful in the preparation of iron oxide nanoparticles with controllable size and high quality. The structure of ferrocene involves equal bonding between all of the ten carbon atoms of the cyclopentadienyl rings and the Fe centre. The stability of ferrocene grows from the strong overlap between the metal d-orbitals and the π-electrons in the p-orbitals of the cyclopentadienyl ligands. Ferrocene, is an important material for preparing iron oxide nanostructures through thermal decomposition, e.g., magnetic micro/nanoparticles [24-28], thin films [29, 30], single walled/ferromagnetic-filled carbon nanotube [31-35], nanocomposites [36]. Taking a leaf out of these works, we undertook a study on the solid state thermal decomposition of ferrocene material in the presence of guest molecules of various kinds leading to iron oxides [37, 38].

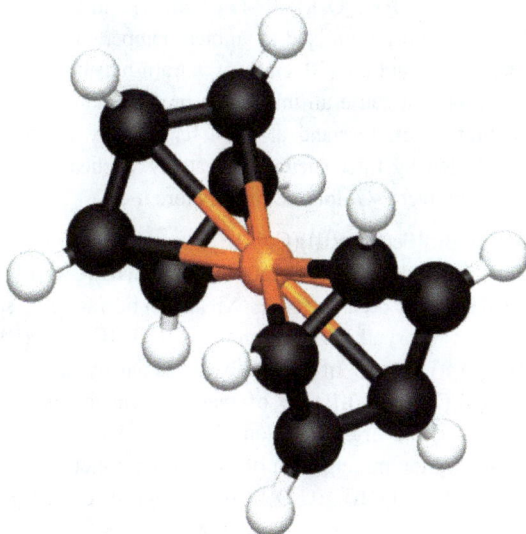

Figure 2. Schematic diagram of ferrocene [(C₅H₅)₂Fe] molecule. (Black spheres: C, Orange sphere: Fe; white spheres: H)

In the following sections, we discuss on the synthesis of nanoparticles using (i) $\{XR_4[M(II)M(III)(C_2O_4)_3]\}_\infty$ and (ii) ferrocene materials as precursors through solid state thermal decomposition route and their characterization mainly by spectroscopic techniques.

2. Materials

2.1 Synthesis of $\{XR_4[M(II)M(III)(C_2O_4)_3]\}_\infty$ type precursors for thermal decomposition

The mixed metal complexes like $\{XR_4[M(II)M(III)(C_2O_4)_3]\}_\infty$ can be prepared in a one pot reaction [16, 17]. Here we provide a synthesis in an attempt to exhibit the simplicity of the synthetic procedure of this class of materials.

To prepare ferromagnetic $NBu_4[Fe(II)Cr(III)(C_2O_4)_3]$ complex, chemicals required are freshly prepared $K_3[Cr(III)(C_2O_4)_3]).3H_2O$, $FeSO_4.7H_2O$ and tetra-*n* butyl ammonium bromide. The total reaction must be carried out in an inert atmosphere. First, an aqueous solution ($\sim$10 cm^3) of $K_3[Cr(III)(C_2O_4)_3]).3H_2O$ (2 mmol) and an aqueous solution ($\sim$10 cm^3) of $FeSO_4.7H_2O$ (2 mmol) are mixed at ambient temperature. Then to the resulting solution another aqueous solution (10 cm^3) of tetra(n-butyl)ammonium bromide (2 mmol) is added drop wise to cause an immediate precipitation of green microcrystals. After this, the solution is left to stand at room temperature for one hour. Then the microcrystals are collected by filter suction, repeatedly washed with de-aerated water, and then dried in vacuum over P_2O_5 for one week before use.

The ferrimagnetic $\{NBu_4[Fe(II)Fe(III)(C_2O_4)_3]\}_\infty$ complex can be synthesized by following the same method only by replacing $K_3[Cr(III)(C_2O_4)_3]).3H_2O$ salt with $K_3[Fe(III)(C_2O_4)_3]).3H_2O$. The organic cation '$XR_4^+$' can be varied by selecting different suitable alkyl/aryl salts to synthesize a range of $\{XR_4[Fe(II)Fe(III)(C_2O_4)_3]\}_\infty$ or $\{XR_4[Fe(II)Cr(III)(C_2O_4)_3]\}_\infty$ type of materials. It is easily understandable that (i) $NBu_4[M_1(II)_xM_2(II)_{1-x}M_3(III)(C_2O_4)_3]$ type of materials can be prepared adopting the same method by keeping the organic cation XR_4^+ and $K_3[M_3(III)(C_2O_4)_3].3H_2O$ salt unchanged but by varying the mixing ratio of the di-valent metal ($M_1(II)$, $M_2(II)$) salts, and (ii) $NBu_4[M_1(II)M_2(III)_xM_3(III)_{1-x}(C_2O_4)_3]$ type materials can be prepared when the organic cation XR_4^+ and the di-valent metal (M_1) salt are kept fixed while the ratio of the trivalent salts - $K_3[M_2(III)(C_2O_4)_3]$ and $K_3[M_3(III)(C_2O_4)_3]$ are varied as desired. The materials were analyzed for their elemental occupancy to determine of the mass fractions of C, H, N and metals. The microanalysis data obtained were found to be in general agreement with the calculated data.

Following this method a large number of precursor materials for solid state thermal decomposition were synthesized. However, Table 1 provides the reaction details for the preparation of some $\{XR_4[M(II)M(III)(C_2O_4)_3]\}_\infty$ type of precursors which will be used for thermal decomposition.

Table 1: Reaction details for the preparation of some $\{XR_4[M(II)M(III)(C_2O_4)_3]\}^1_\infty$ type of precursors used for thermal decomposition.

Desired precursor	1st aqueous solution	2nd aqueous solution	3rd aqueous solution	Nature/ colour of the precipitates
$\{N(n\text{-}C_4H_9)_4[Fe(II)Fe(III)(C_2O_4)_3]\}_\infty$	$N(n\text{-}C_4H_9)_4Br$	$K_3[Fe(C_2O_4)_3].3H_2O$	$FeSO_4.7H_2O$	dark green microcrystals
$\{N(n\text{-}C_4H_9)_4[Co(II)Fe(III)(C_2O_4)_3]\}_\infty$	$N(n\text{-}C_4H_9)_4Br$	$K_3[Fe(C_2O_4)_3].3H_2O$	$Co(NO_3)_2.6H_2O$	brick-red microcrystals
$\{As(C_6H_5)_4[Fe(II)Fe(III)(C_2O_4)_3]\}_\infty$	$As(C_6H_5)_4Br$	$K_3[Fe(C_2O_4)_3].3H_2O$	$FeSO_4.7H_2O$	pale green microcrystals
$\{N(n\text{-}C_5H_{11})_4[Mn(II)Fe(III)(C_2O_4)_3]\}_\infty$	$N(n\text{-}C_5H_{11})_4I$	$K_3[Fe(C_2O_4)_3].3H_2O$	$MnCl_2.4H_2O$	pale green microcrystals
$\{N(n\text{-}C_4H_9)_4[Fe(II)Cr(III)(C_2O_4)_3]\}_\infty$	$N(n\text{-}C_4H_9)_4Br$	$K_3[Cr(C_2O_4)_3].3H_2O$	$FeSO_4.7H_2O$	dark brown microcrystals

2.2 Synthesis of ferrocene-based precursors for thermal decomposition

The ferrocene material undergoes cent percent thermal decomposition/sublimation at ~175 °C. It is important to explore this phenomenon in order to produce iron oxide nano/microparticles utilizing ferrocene. Thermal decomposition of ferrocene is studied solid reaction environments. Precursors were prepared by mixing ferrocene with selected organic compounds in different weight ratios. The organic compounds used with ferrocene were (i) Oxalic acid dihydrate[$(COOH)_2.2H_2O$] and (ii) Fe(II) acetylacetonate($C_{10}H_{14}FeO_4$). The schematic diagrams of these two organic compounds are shown in Fig. 3.

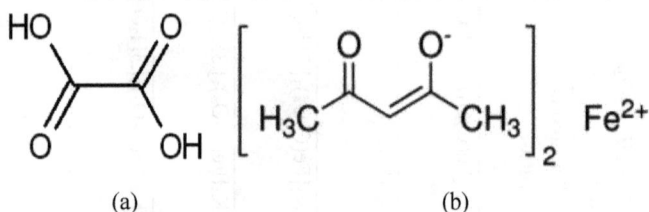

(a)   (b)

Figure. 3. Schematic diagram of (a) Oxalic acid dihydrate (two water H_2O molecules are not shown) and (b) Fe(II)acetylacetonate.

The preparation of individual precursors is discussed below:

- *Solid mixtures of ferrocene and oxalic acid dihydrate*

Solid mixtures of polycrystalline ferrocene and oxalic acid dihydrateare were prepared by physical mixing using mortar and pestle in two different weight ratios - (3:2, 2:3). The corresponding samples are termed as A1 and A2, respectively. The decomposed samples are termed as A1D and A2D accordingly.

- *Solid mixtures of ferrocene and Fe(II)acetylacetonate*

A solid mixture was prepared by grinding high quality polycrystalline ferrocene and Fe(II)acetylacetonate together in 2:3 weight ratio using a mortal-pestle. The samples to study are termed as B1 and the corresponding thermally decomposed material is termed as B1D.

2.3 Synthesis of thermally decomposed material

A convenient amount of the microcrystalline precursors is kept in a porcelain boat and then loaded into an indigenous programmable laboratory furnace at a pre-determined temperature for 7-8 hours in ambient atmosphere for thermal decomposition. After the decomposition the sample material was left for cooling inside the furnace and then collected at room temperature. Accordingly, the various metal oxides of interesting facets are obtained.

2.4   Characterization

The decomposed materials were then characterized by various physical techniques like SEM/TEM, EDAX, powder X-ray diffraction (XRD), SQUID, Mössbauer spectroscopy. However, in the following sections discussion on the SEM/TEM images obtained will be majorly made to describe the morphological nature of the particles grown by solid state thermal decomposition of molecular organometallic materials.

3.   Results

3.1   Nanomaterials obtained from $\{XR_4[M(II)M(III)(C_2O_4)_3]\}_\infty$ type precursors

3.1.1 $\{N(n\text{-}C_4H_9)_4[M(II)Fe(III)(C_2O_4)_3]\}_\infty$, M = Zn, Co, Fe

XRD pattern of the thermally degraded samples of $\{N(n\text{-}C_4H_9)_4[M(II)Fe(III)(C_2O_4)_3]\}_\infty$, M = Zn, Co, Fe (say, BuZnFe-D, BuCoFe-D and BuFeFe-D) was obtained using Cu-K$_\alpha$ line within $10^0 < 2\theta < 140^0$ range. From the XRD peak broadening the mean crystallite size (D) was estimated using Scherer formula. The XRD pattern of all these materials exhibit that the diffraction peaks/lines correspond to iron and zinc oxide in case of BuZnFe-D, iron and cobalt oxide in case of BuCoFe-D and hematite for BuFeFe-D [19]. The structure, unit cell parameters and mean crystallite size (D) of these materials are presented in Table 2. The XRD observations establish that nano-sized ferrite particles can be obtained by thermal decomposition of the molecular precursors $\{N(n\text{-}C_4H_9)_4[M(II)Fe(III)(C_2O_4)_3]\}_\infty$.

Surface morphology and particle nature of the thermally decomposed material BuZnFe-D, BuCoFe-D and BuFeFe-D were observed by SEM. Fig. 4 - Fig. 6 represent the SEM pictures of the decomposed materials. These Figs. clearly show that nanoparticles are formed on thermal decomposition of BuZnFe-O, BuCoFe-O and BuFeFe-O. In case of BuZnFe-D, the nano-particles linked in a chain accompanied by agglomeration, whereas the nano-particles appeared in flakes of irregular geometry with cloudy boarder in case of BuCoFe-D. As observed through SEM picture of BuFeFe-D, nanoparticles show rod like structure of more or less uniform diameter and sharp geometry and the agglomeration of tiny rods bears length ranging between 580 and 210 nm and diameter between 210 and 110 nm [19].

Table 2. XRD pattern of the thermal decomposed materials of {N(n-C₄H₉)₄[M(II)Fe(III)(C₂O₄)₃]}∞ with M = Zn, Co and Fe.

Decomposed Material	Structure	Unit cell parameter	Mean crystallite size
BuZnFe-**D**	Maghemite 52.90%	$a = b = c = 8.43652$ Å $\alpha = \beta = \gamma = 90.0°$	45 nm
	Zincite 47.10%	$a = b = 3.24892$ Å $c = 5.20422$ Å $\alpha = \beta = \gamma = 90.0°$	65 nm
BuCoFe-**D**	Maghemite 72%	$a = b = c = 8.36229$Å $\alpha = \beta = \gamma = 90.0°$	68 nm
	Cobalt Oxide 28%	$a = 8.12440$Å $b = 5.91274$ Å $c = 5.12243$ Å $\alpha = \beta = \gamma = 90.0°$	19 nm
BuFeFe-**D**	Hematite, α-Fe₂O₃	$a = b = 5.03286$ Å $c = 13.74430$ Å $\alpha = \beta = \gamma = 90.0°$	169 nm

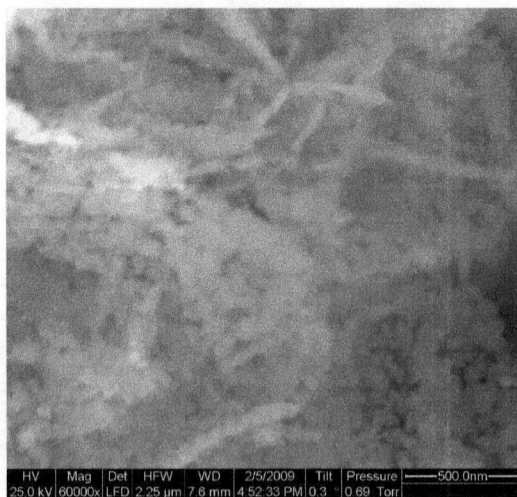

Figure. 4. SEM micrograph of the thermally decomposed material BuZnFe-D : Nano-particles are linked in a chain accompanied by agglomeration.

Figure. 5. SEM micrograph of the thermally decomposed material BuCoFe-D: Nano-particles appeared in flakes with irregular geometry with cloudy boarder.

Figure. 6. SEM micrograph of the thermally decomposed material BuFeFe-D: Nanoparticles show rod like structure of more or less uniform diameter and sharp geometry.

3.1.2 $\{N(n\text{-}C_5H_{11})_4[Mn(II)Fe(III)(C_2O_4)_3]\}_\infty$

XRD pattern of the thermally degraded $\{N(n\text{-}C_5H_{11})_4[Mn(II)Fe(III)(C_2O_4)_3]\}_\infty$ (say, PtMnFe-D) was obtained by following method as described earlier. The XRD pattern exhibits that the diffraction peaks correspond to wustite-like phase presumably of the (Fe, Mn)$_{1-z}$ O) type. The unit cell parameters are a = 6.65439 Å, b = 4.70554 Å, c = 3.32508 Å, $\alpha = \beta = \gamma = 90.0°$. The mean crystallite size (D) of this material is 35.23nm. Thus, XRD study establishes that nano-sized wustite-like particles can be obtained by thermal decomposition of the molecular precursors $\{N(n\text{-}C_5H_{11})_4[Mn(II)Fe(III)(C_2O_4)_3]\}_\infty$.

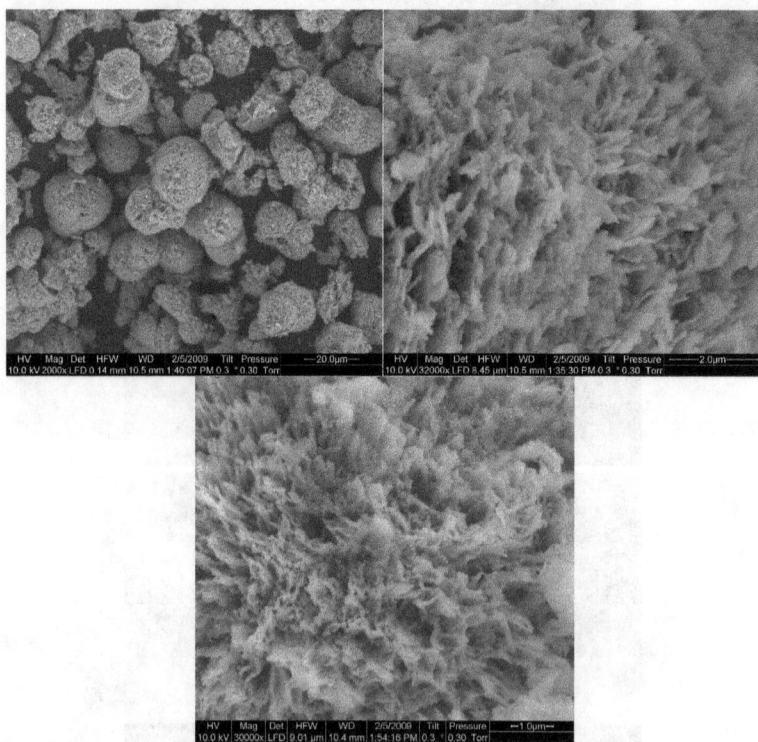

Figure. 7. SEM micrographs of the thermally decomposed material PtMnFe-D at different magnifications: Nanoparticles appear in puffy ball-shapes - surface of these shapes is of leafy structure - all extended outwards from centre and leaves are very thin lying in nano scale.

Morphology and particle nature of PtMnFe-**D** was studied by SEM. Figure 6 represents the SEM micrographs of PtMnFe-**D** at different magnifications. This Figure clearly shows that the decomposed materials are available in puffy ball-shapes. The surface of these shapes is of leafy structure - all extended outwards from the centre. These leaves are very thin and obviously within nanometer scale.

3.1.3 {As(C₆H₅)₄[Fe(II)Fe(III)(C₂O₄)₃]}∞

Powder X-ray diffraction pattern of thermally decomposed $\{As(C_6H_5)_4[Fe(II)Fe(III)(C_2O_4)_3]\}_\infty$ (say, AsFeFe-D) shows that all diffraction peaks correspond to the hematite structure with unit cell parameters $a = b = 5.03571$ Å; $c = 13.74985$ Å, $\alpha = \beta = \gamma = 90.0°$ and the mean crystallite size $D = 34.7(0.04)$ μm. Particle nature of AsFeFe-D was further verified by transmission electron microscopy (TEM) as shown in Fig. 8.

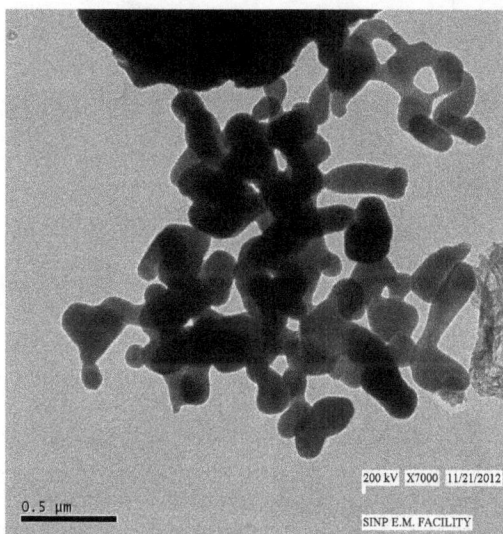

Figure 8. TEM micrograph of the thermally decomposed material AsFeFe-D.

3.1.4 $\{N(n-C_4H_9)_4[Fe(II)Cr(III)(C_2O_4)_3]\}_\infty$

Figure. 9. SEM micrograph of the thermally decomposed material BuFeCr-D.

3.1.5 Ferrocene with Oxalic acid dihydrate

The X-ray powder diffraction patterns of the thermally decomposed samples –A1d and A2d exhibit that all diffraction peaks correspond to hematite with unit cell parameters: $a = b = 5.0322(4)$ Å; $c = 13.7460(13)$ Å, $\alpha = \beta = 90.0°$, $\gamma = 120.0°$ for A1d, and $a = b = 5.0322(2)$ Å; $c = 13.7441(8)$ Å, $\alpha = \beta = 90.0°$, $\gamma = 120.0°$ for A2d. From the XRD peak broadening the mean crystallite size (D) was estimated 41 nm and 37 nm for A1d and A2d, respectively. This result establishes that nanosized hematite particles are formed on thermal decomposition of ferrocene in the presence of oxalic acid. The morphology of A1D and A2D was observed by SEM. Fig. 10 represents the SEM micrographs of A1D and A2D. From these pictures it is clear that the decomposed materials are the agglomerations of tiny particles of various shapes and sizes. The material A1D largely contains particles with sharp edges, whereas the particles in A2D are in general round-shaped.

Figure 10. SEM micrographs of the decomposed materials: (a) A1D and (b) A2D.

3.1.6 Ferrocene with Fe(II)acetylacetonate

The structure and composition of the decomposed material was investigated by the powder XRD method. XRD pattern of B1D is analyzed to identify the components, unit cell parameters and mean crystallite size of the material. The estimated lattice parameters for B1d are $a = b = 5.0322$ Å; $c = 13.7378$ Å, $\alpha = \beta = 90.0°$, $\gamma = 120°$. The XRD pattern clearly exhibits that all diffraction peaks / lines correspond to hematite. The mean crystallite size D was estimated = 260 nm. This result establishes that large-sized hematite particles can be synthesized on thermal decomposition of ferrocene in the presence of Fe(II)acetylacetonate.

Surface morphology and particle nature of B1D were observed by SEM (Fig. 11). It indicates that the material formed on thermal decomposition is agglomerated particle in the form of tiny rod. There is a variation in the length (250 – 350 nm) and breadth (100 -

150 nm) of the tiny rods as demonstrated by SEM. The observed mean size of B1D is close to the mean crystallite size estimated through the powder XRD study.

Figure 11. SEM micrograph of the decomposed material B1D.

4.   Applications for biological environment

From the perspective of environmental surveillance to protect living organism monitoring of major polluting toxic gases like NH_3, CO, NO, etc. at source points is an important consideration since exposure of such toxic gases to living habitat exceeding minimum permissible limits brings about detrimental changes to the biological environment. There are quite a few reports that prove beyond doubts that mainly double-walled carbon nano-tubes when attached with nano-oxides through proper functionalisation sense toxic gases of even bbm level. For example, the design and fabrication of NH_3 gas sensor making use of SWNT (single-walled carbon nanotube) / MWCNT (multi-walled carbon nanotube) as template and metal oxide or metal nanocomposite as decorating material has drawn attention of various research groups across the globe.

Magnetic iron oxide nanoparticles coated with suitable shells e.g. kappa-carrageenan, silica can be employed as adsorbents for removal of methylene blue (MB) from water since the discharge of effluents contain such dyes extensively used in the industry having harmful effects in the environment [39]. A few metal oxides like ZnO in their nano structures are found as promising candidates in bio-sensing, drug delivery and biomedical imaging for their photoluminescence attribute [40]. Nanomaterials (NMs) are increasingly found in biotechnology, biomedicine and environmental field and for example, some biomolecules such as proteins rapidly bind to NMs creating biomolecule corona that controls the *in vitro* and/or *in vivo* NM uses in humans and ecosystems too. It is a sterling revelation that ferrite magnetic nanoparticles (MNPs) can be used as heating mediators for magnetic hyperthermia and show promises to be used for cancer treatment. It is reported [41] that the highly crystalline MNPs demonstrated an exceptional magnetic heating during hyperthermia. The group as in [41], has carried out *in vitro* application on the human osteosarcoma cell line Saos-2 incubated with manganese ferrite nanoparticles and paved the ways for magnetic hyperthermia driven applications. Magnetic resonance imaging (MRI) has emerged as an alternative technique for early detection and treatment of many dreaded diseases like cancer, neurodegenerative disorders, vascular ailments, etc. The inherent shortcomings with respect to sensitivity and specificity of conventional MRI are astounded by new kind of contrast agents that exhibits improved magnetic and biological properties in nanoscale. The group of researchers, Leal and others [42] develop contrast agent based on manganese ferrite nanoparticles (MNPs) with size varying from 6 to 14 nm and form PEG (polyethylene glycol)-ylated MNPs which demonstrate different relaxivities depending on their size and the magnetic field applied. These polymer core shells of the PEGylated MNPs reduce cytotoxicity markedly, and permit long blood circulation time and under low magnetic field they make these MNs very promising contrast agents for MRI.

References

[1]   N.R. Jana, L. Gearheart, C.J. Murphy, 'Seed-Mediated Growth Approach for Shape-Controlled Synthesis of Spheroidal and Rod-like Gold Nanoparticles Using a Surfactant Template', Adv. Mater.13 (2001) 1389-1393. https://doi.org/10.1002/1521-4095(200109)13:18<1389::AID-ADMA1389>3.0.CO;2-F

[2]   J. Gao, C.M. Bender, C.J. Murphy, 'Dependence of the Gold Nanorod Aspect Ratio on the Nature of the Directing Surfactant in Aqueous Solution', Langmuir 19 (2003) 9065-9070. https://doi.org/10.1021/la034919i

[3]   A. Gole, C.J. Murphy, 'Seed-Mediated Synthesis of Gold Nanorods: Role of the Size and Nature of the Seed', Chem. Mater. 16 (2004) 3633-3640. https://doi.org/10.1021/cm0492336

[4]   T.K. Sau, C.J. Murphy, Room Temperature, 'High-Yield Synthesis of Multiple Shapes of Gold Nanoparticles in Aqueous Solution', J. Am. Chem. Soc. 126 (2004) 8648-8649. https://doi.org/10.1021/ja047846d

[5]   Y. Wang, Y. Zheng, C.Z. Huang, Y. Xia, 'Synthesis of Ag Nanocubes 18–32 nm in Edge Length: The Effects of Polyol on Reduction Kinetics, Size Control, and Reproducibility', J. Am. Chem. Soc. 135 (2013) 1941-1951. https://doi.org/10.1021/ja311503q

[6]   Y. Sun, Y. Xia, 'Shape-Controlled Synthesis of Gold and Silver Nanoparticles', Science 298 (2002) 2176-2179. https://doi.org/10.1126/science.1077229

[7]   S.H. Im, Y.T. Lee, B. Wiley, Y. Xia,' Large-Scale Synthesis of Silver Nanocubes: The Role of HCl in Promoting Cube Perfection and Monodispersity', Angew. Chem., Int. Ed. 44 (2005) 2154-2157. https://doi.org/10.1002/anie.200462208

[8]   P. Harmankova, M. Hermanek, R. Zboril, 'Thermal Decomposition of Ferric Oxalate Tetrahydrate in Oxidative and Inert Atmospheres: The Role of Ferrous Oxalate as an Intermediate', Eur. J. Inorg. Chem. (2010) 1110-1118. https://doi.org/10.1002/ejic.200900835

[9]   Y.C. Zhang, J.Y. Tang, X.Y. Hu, 'Controllable Synthesis and Magnetic Properties of Pure Hematite and Maghemite Nanocrystals from a Molecular Precursor', J. Alloys. Compd. 462 (2008) 24-28. https://doi.org/10.1016/j.jallcom.2007.07.115

[10]   H.H. Kung, Transition Metal Oxides: Surface Chemistry and Catalysis, in : B. Delmon, J. T. Yates (Eds.), Studies in Surface Science and Catalysis, 1st Ed., Vol. 45, Elsevier, Amsterdam, The Netherlands, 1989.

[11]   N. Pinna, G. Garnweitner, M. Antonietti, M. Niederberger,'A General Nonaqueous Route to Binary Metal Oxide Nanocrystals Involving a C−C Bond Cleavage', J. Am. Chem. Soc. 127 (2005) 5608-5612. https://doi.org/10.1021/ja042323r

[12]   S. Chaianansutcharit, O. Mekasuwandumrong, P. Praserthdam, 'Synthesis of Fe_2O_3 Nanoparticles in Different Reaction Media', Ceram. Inter. 33 (2007) 697-699. https://doi.org/10.1016/j.ceramint.2005.12.013

[13]   Z.X. Tang, S. Nafis, C.M. Sorensen, G.C. Hadjipanayis, K.J. Klabunde, 'Magnetic Properties of Aerosol Synthesized Iron Oxide Particles', J. Magn. Magn. Mater. 80 (1989) 285-289. https://doi.org/10.1016/0304-8853(89)90131-5

[14] S.J. Campbell, W.A. Kaczmarek, G.M. Wang, 'Mechanochemical Transformation of Hematite to Magnetite', Nanostructured. Mater. 6 (1995) 735. https://doi.org/10.1016/0965-9773(95)00163-8

[15] A. Bhattacharjee, 'A Legendary Molecular Magnetic System: A[M(II)M(III)(C$_2$O$_4$)$_3$]', Curr. Inorg. Chem. 6 (2017) 162-180. https://doi.org/10.2174/1877944107666161208120622

[16] H. Tamaki, J. Zhong, N. Matsumoto, S. Ida, M. Koikawa, N. Achiwa, Y. Hashimoto, H. Ōkawa, 'Design of Metal-Complex Magnets. Syntheses and Magnetic Properties of Mixed-Metal Assemblies {NBu$_4$[MCr(ox)$_3$]}$_\infty$ (NBu^{4+} = tetra(n-butyl)ammonium ion; ox^{2-} = oxalate ion; M = Mn^{2+}, Fe^{2+}, Co^{2+}, Ni^{2+}, Cu^{2+}, Zn^{2+})', J. Am. Chem. Soc. 114 (1992) 6974-6979. https://doi.org/10.1021/ja00044a004

[17] H. Ōkawa, N. Matsumoto, H. Tamaki, M. Ohba, 'Ferrimagnetic Mixed-Metal Assemblies {NBu$_4$[MFe(ox)$_3$]}$_\infty$', Mol. Cryst. Liq. Cryst. 233 (1993) 257-262. https://doi.org/10.1080/10587259308054965

[18] K.E. Neo, Y.Y. Ong, H.V. Huynh, T.S. Andy Horr, 'A Single-molecular Pathway from Heterometallic MM' (M = Ba^{II}, Mn^{II}; M' = Cr^{III}) Oxalato Complexes to Intermetallic Composite Oxides', J. Mater. Chem. 17 (2007) 1002-1006. https://doi.org/10.1039/B609630A

[19] A. Bhattacharjee, D. Roy, M. Roy, S. Chakraborty, A. De, J. Kusz, W. Hofmeister, 'Rod-like Ferrites Obtained through Thermal Degradation of a Molecular Ferrimagnet', J. Alloy. Comps. 503 (2010) 449-453. https://doi.org/10.1016/j.jallcom.2010.05.031

[20] D. Roy, M. Roy, M. Zubko, J. Kusz, A. Bhattacharjee, 'Solid-State Thermal Reaction of a Molecular Material and Solventless Synthesis of Iron Oxide', Int. J. Thermophys. 37 (2016) 93-108. https://doi.org/10.1007/s10765-016-2099-0

[21] A. Bhattacharjee, D. Roy, M. Roy, 'Thermal Degradation of a Molecular Magnetic Material: {N(n-C$_4$H$_9$)$_4$[Fe^{II}Fe^{III}(C$_2$O$_4$)$_3$]}$_\infty$', J. Therm. Anal. Cal. 109 (2012) 1423-1427. https://doi.org/10.1007/s10973-011-1829-6

[22] A. Bhattacharjee, D. Roy, M. Roy, 'Thermal Decomposition of Molecular Materials {N(n-C$_4$H$_9$)$_4$[M^{II}Fe^{III}(C$_2$O$_4$)$_3$]}$_\infty$,M^{II} = Zn, Co, Fe', Int. J. Thermophys. 33 (2012) 2351-2365. https://doi.org/10.1007/s10765-012-1293-y

[23]   D. Roy, Study of thermal decomposition of some oxalate-based molecular
        materials leading to metal oxides (Ph.D. Thesis, Visva-Bharati University, India,
        2013).

[24]   G. Huang, J. Weng, 'Syntheses of Carbon Nanomaterials by Ferrocene', Curr.
        Org. Chem. 15 (2011) 3653-3666. https://doi.org/10.2174/138527211797884593

[25]   N. Koprinarov, M. Konstantinova, M. Marinov, 'Ferromagnetic Nanomaterials
        Obtained by Thermal Decomposition of Ferrocene', Solid State Phenomena 159
        (2010) 105-108. https://doi.org/10.4028/www.scientific.net/SSP.159.105

[26]   K. Elihn, L. Landström, O. Alm, M. Boman, P. Heszler, 'Size and Structure of
        Nanoparticles formed via Ultraviolet Photolysis of Ferrocene', J. Appl. Phys. 101
        (2007) 34311-34315. https://doi.org/10.1063/1.2432406

[27]   E.P. Sajitha, V. Prasad, S.V. Subramanyam, A.K. Mishra, S. Sarkar, C. Bansal,
        'Structural, Magnetic and Mössbauer Studies of Iron Inclusions in a Carbon
        Matrix', J. Magn. Magn. Mater. 313 (2007) 329-336.
        https://doi.org/10.1016/j.jmmm.2007.02.001

[28]   D. Amara, J. Grinblat, S. Margel, 'Solventless Thermal
        Decomposition of Ferrocene as a New Approach for One-Step Synthesis
        of Magnetite Nanocubes and Nanospheres', J. Mater. Chem. 22 (2012) 2188-2195.
        https://doi.org/10.1039/C1JM13942H

[29]   S. Saremi-Yarahmadi, A.A. Tahir, B. Vaidyanathan, K.G.U. Wijayantha,
        'Fabrication of Nanostructured α-Fe_2O_3 Electrodes using Ferrocene for Solar
        Hydrogen Generation', Mater. Lett. 63 (2009) 523-526.
        https://doi.org/10.1016/j.matlet.2008.11.011

[30]   M. Rooth, A. Johansson, M. Boman, A. Hårsta, 'Atomic Layer Deposition of Iron
        Oxide Thin Films and Nanotubes using Ferrocene and Oxygen as Precursors',
        Chem. Vap. Deposition 14 (2008) 67-70. https://doi.org/10.1002/cvde.200706649

[31]   R. Shah, X.F. Zhang, X. An, S. Kar, S. Talapatra, 'Ferrocene Derived Carbon
        Nanotubes and Their Application as Electrochemical Double Layer Capacitor
        Electrodes', J. Nanosci. Nanotechnol. 10 (2010) 4043-4048.
        https://doi.org/10.1166/jnn.2010.2406

[32]   A.G. Nasibulin, S.G. Shandakov, A.S. Anisimov, D. Gonzalez, H. Jiang, M.
        Pudas, P. Queipo, E.I. Kauppinen, 'Charging of Aerosol Products during
        Ferrocene Vapor Decomposition in N2 and CO Atmospheres', J. Phys. Chem. 112
        (2008) 5762-5769. https://doi.org/10.1021/jp0755995

[33] A. Barreiro, S. Hampel, M.H. Rümmeli, C.K. Kramberger, A. Grüneis, K. Biedermann, A. Leonhardt, T. Gemming, B. Büchner, A. Bachtold, T. Pichler, 'Thermal Decomposition of Ferrocene as a Method for Production of Single-Walled Carbon Nanotubes without Additional Carbon Sources', J. Phys. Chem. B 110 (2006) 20973-20977. https://doi.org/10.1021/jp0636571

[34] K.E. Kim, K.J. Kim, W.S. Jung, S.Y. Bae, J. Park, J. Choi, J. Choo, 'Investigation on the Temperature-Dependent Growth Rate of Carbon Nanotubes using Chemical Vapor Deposition of Ferrocene and Acetylene', Chem. Phys. Lett. 401 (2005) 459-464. https://doi.org/10.1016/j.cplett.2004.11.113

[35] A. Leonhardt, S. Hampel, C. Muller, I. Mönch, R. Koseva, M. Ritschel, D. Elefant, K. Biedermann, B. Büchner, 'Synthesis, Properties, and Applications of Ferromagnetic-Filled Carbon Nanotubes', Chem. Vap. Deposition 12 (2006) 380-387. https://doi.org/10.1002/cvde.200506441

[36] R. Prakash, A.K. Mishra, A. Roth, C. Kübel, T. Scherer, M. Ghafari, H. Hahn, M. Fichtner, 'A Ferrocene-Based Carbon–Iron Lithium Fluoride Nanocomposite as a Stable Electrode Material an Lithium Batteries', J. Mater. Chem. 20 (2010) 1871-1876. https://doi.org/10.1039/b919097j

[37] A. Bhattacharjee, A. Rooj, M. Roy, J. Kusz, P. Gütlich, 'Solventless Synthesis of Hematite Nanoparticles using Ferrocene', J. Mater. Sci. 48 (2013) 2961-2968. https://doi.org/10.1007/s10853-012-7067-x

[38] A. Rooj, M. Roy, J. Kusz, A. Bhattacharjee, 'A Solventless Method to Prepare Hematite using Thermal Decomposition of Ferrocene', Int. J. Exp. Spect. Tech. 1 (2016) 3-10.

[39] S.F. Soares, T.R. Simoes, T. Trindade, A.L. Daniel-da-Silva, 'Highly Efficient Removal of Dye from Water Using Magnetic Carrageenan/Silica Hybrid Nano-adsorbents', Water Air & Soil Pollut. 228 (2017) 87-98. https://doi.org/10.1007/s11270-017-3281-0

[40] M. Ghosh, S. Ghosh, M. Seibt, 'Designing Deoxidation Inhibiting Encapsulation of Metal Oxide Nanostructures for Fluidic and Biological Application', Apl. Surf. Sc. 390 (2016) 924-928. https://doi.org/10.1016/j.apsusc.2016.08.117

[41] A. Makridis, I. Chatzitheodorou, K. Topouridou, M.P. Yavropoulou, M. Angelakeris, C. Dendrinou-Samara, 'A Facile Microwave Synthetic Route for Ferrite Nanoparticles with Direct Impact in Magnetic Particle Hyperthermia', Mater. Sci. Eng. C- Mater. Bio. Appl. 63 (2016) 663-670. https://doi.org/10.1016/j.msec.2016.03.033

[42] M.P. Leal, S. Rivera-Fernandez, J.M. Franco, D. Pozo, J.M. de la Fuente, M.L. Garcia-Martin, 'Long-Circulating Pegylated Manganese Ferrite Nanoparticles for MRI-Based Molecular Imaging', Nanoscale 7 (2015) 2050-2059. https://doi.org/10.1039/C4NR05781C

Methodology of Physical Characterization

Instrumental methods, use of state-of-the-art techniques for knowing and deriving the information on engineered nano-particles/ nano-scaled materials is special branch of study by itself.

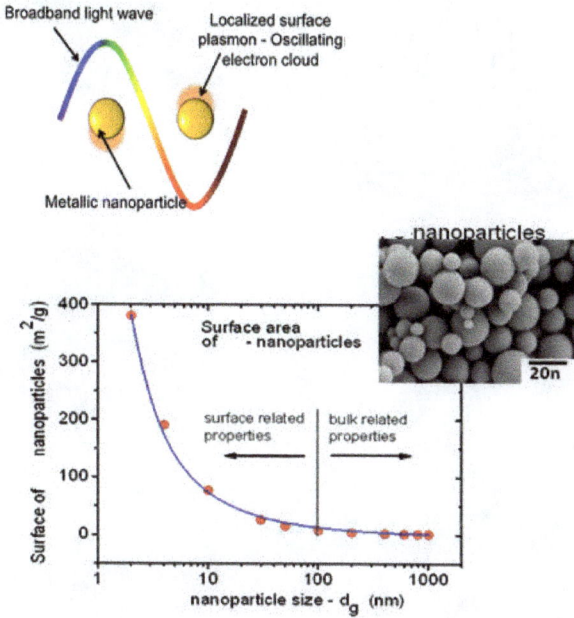

Chapter 4

Methodology and Physical Characterization of Nanoparticles using Photophysical Techniques

Arnab Maity[1] and Samita Basu[2*]

[1]Department of Chemistry, Akal University, Talwandi Sabo, Bhatinda, Punjab-15302 [2]Chemical Sciences Division, Saha Institute of Nuclear Physics, 1/AF Bidhannager Kolkata, India

*samita.basu@saha.ac.in

—"Old Man's Advice to Youth: 'Never Lose a Holy Curiosity.'" LIFE Magazine (May 2nd, 1955) p. 64" — Albert Einstein

Abstract

Photochemical techniques involving UV-VIS absorption and emission of light quanta have yielded significant information through electronic perturbation of surface electrons in nano-particle systems used for biological applications. Elucidation of various mechanistic processes improvised the use of sophisticated photonic detecting systems that have been discussed in a simple and lucid manner in this article through citation of literature examples and also from the works of authors' own laboratory. This includes steady-state absorption and emission spectroscopy, time-resolved photodynamic spectroscopy, laser flash photolysis study as well as fluorescence photo imaging systems applied to cellular applications.

Keywords

UV-Vis Absorption Spectroscopy, Plasmon Absorption Band, Ru:CNDEDAs, Photo induced Electron Transfer (PET), TCSPC, Femtosecond Fluorescence Upconversion, Laser Flash Photolysis

Contents

1. Introduction..76

2. UV-Vis absorption spectroscopy...77

3. Steady state fluorescence spectroscopy and imaging technique 81

    3.1 Fluorescence imaging technique 84

    3.2 Time resolved fluorescence spectroscopic technique......................87

4. Femtosecond fluorescence upconversion studies................................91

5. Laser flash photolysis 95

5. Conclusion 98

References...99

1. Introduction

For the last few decades considerable interest have grown in the field of nanotechnology involving metals, semiconductors, non-metals, metal oxides as well as organic molecular assemblies as nanoparticles. A wide variety of applications in the field of information technology, biomedical, energy storage, catalysis, bio-imaging, and also environmental studies have been established utilising the unique properties determined primarily by their size, composition and structures along with their ability to form self organised assemblies. In this chapter, we discuss various characterization techniques which are frequently used to study nanoparticles and their ultimate fate in the respective field. Optical spectroscopic techniques offer a unique platform to determine various properties *viz* shape, size, microenvironment, kinetic activities, energy transfer phenomena, etc. with practical ease and efficiency since electronic structures can be easily perturbed through absorption of electromagnetic radiations usually in the UV-Vis region (~200-800 nm or 1.55-6.20 eV) and in the near infrared region (~800-1100 nm or 1.13-1.55 eV). In this article, various photochemical techniques those usually employed in describing the phenomenology of nanoparticles in solution phase have been clearly delineated in separate paragraphs. Fig .1. summerizes the various techniques involved in this article.

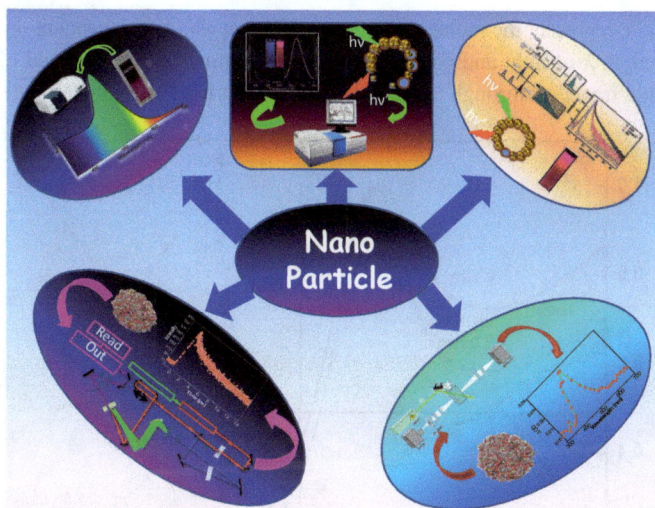

Figure 1.   The various photon absorption - emission processes in the UV-VIS region and their experimental analyses in steady state as well as in photo-kinetic state have been captured pictorially. From top left clockwise: a) UV-Vis absorption spectroscopic technique, b) steady-state fluorescence spectroscopic technique, c) time-correlated-single-photon counting technique, d) laser flash photolysis technique and e) femtosecond fluorescence upconversion technique.

2.   UV-Vis absorption spectroscopy

When a wavelength of electromagnetic radiation, light (λ) of a particular intensity (I_0) falls on a sample, a definite fraction of the light is absorbed by the sample and the rest of the light (I_t) is transmitted through it. This phenomenon is quantified by a parameter which is termed as absorbance (A) or optical density (O.D.) and represented as:

$$A \text{ or O.D.} = \log_{10}(I_0/I_t). \tag{i}$$

The absorption spectrum for that particular sample is obtained by measuring absorbance at varied wavelength. Moreover according to Lambert Beer's law A or O.D. can be quantified by the following equation

$$A \text{ or O.D.} = \varepsilon_v c\, l, \tag{ii}$$

where ε_v and c represent the molar extinction coefficient which is dependent on frequency v or wavelength λ of light and concentration of the sample respectively and l is the path length of light traversed through the sample, which is normally kept as 1 cm.

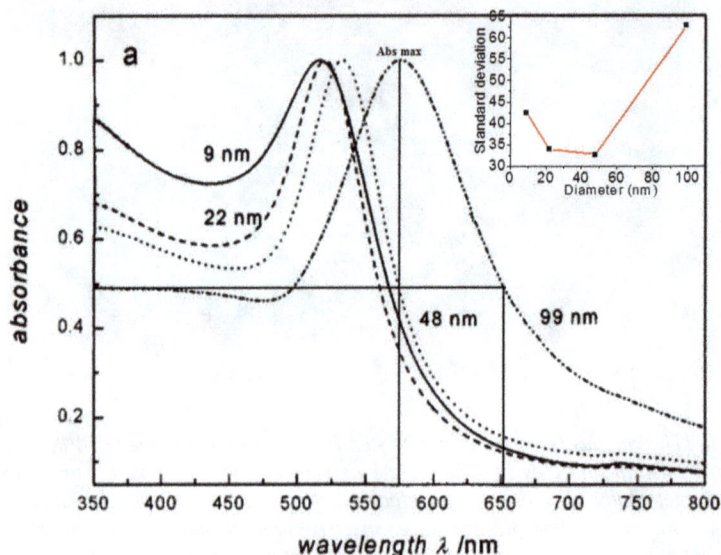

Figure 2. UV-Vis absorption spectra of 5nm, 22 nm, 48nm, 99 nm gold nano particles in water as depicted in ref. 2 and inset shows the change in standard deviations of λ_{max} with particle diameters. The LHS of the respective absorption spectra were symmetrically modified assuming a Gaussian profile and their FWHM were extracted for the calculation of the standard deviation as a parameter. One such illustration has been demonstrated in this figure.

Therefore, since A or O.D. is controlled mainly by ε_v the absorption spectrum should be different for different sample with similar concentration. The wavelength at which the maximum absorbance is obtained is termed as λ_{max}. Now-a-days the UV/Vis/NIR absorption spectroscopy has become one of the simple techniques that is successfully applied to quantify the amount of light absorbed or scattered by the sample. Nanoparticles have optical properties those are sensitive to size, shape, agglomeration state, anti–coagulating or protecting agent, etc. and make UV/Vis/NIR absorption spectroscopy a valuable tool for identifying, quantifying, characterizing and studying materials for a particular reaction process. Nanoparticles (NP) whether it is metal, non metal or semiconductor have an individual characteristic absorption band. For example, N. Kometani et al. [1] studied silver (Ag) and gold (Au) nanoparticles (NP) coated with a dye and reported their characteristic λ_{max} at 390 nm and 520 nm for AgNP and AuNP respectively. M.A. El-Sayed et al. [2] reported that as the size of the Au nanoparticles increased, the absorption maximum (λ_{max}) also shifted to the longer wavelengths as shown in Fig. 2.

From Fig. 2, it is clear that as the size of the Au nanoparticles increases, the λ_{max} is shifted to the longer wavelength along with a change in the line shape of the absorption band. If the absorption bands are considered as Gaussian profiles, one may calculate the standard deviation from the following equation:

$$\Gamma = 2.355 \, \sigma, \tag{iii}$$

where Γ is the full width at half maximum (FWHM) and σ is the standard deviation. Inset of Fig. 2 depicts the variation of standard deviation with increasing diameter of Au nanoparticles. Standard deviation decreases with increase in size of the nanoparticles from 9 nm to 22 nm (diameter) then reaches a minimum value at 48 nm (diameter) particle size, after that again increases as the size of Au- nanoparticle increases from 48 nm to 99 nm (diameter). The optical properties of small particle samples are determined mainly by two contributions namely: the properties of the particles acting as well-isolated individuals and the collective properties of the whole ensemble. Accordingly, the narrow size distributions of (roughly) spherical noble metal particles, within a homogeneous medium, can be considered for describing their plasmon absorption maximum (λ_{max}) and the corresponding band width, Γ. Interestingly, it has been shown that there is a considerable shift of λ_{max} (both blue and red shifts) with decrease and also increase of the size of Ag and Au- nano-particles [3].

The width of the plasmon band (which is FWHM) has been related to the structural effects to which the surface electrons are related. It includes the energy and momentum of those electrons engaged in photon absorption, dielectric effect of the particles and subtle structural effects like grain boundaries, impurities, dislocations, etc. of the Au- nanoparticles.

Ru doped carbon nanoparticles have been synthesized from citric acid by pyrolysis method and subsequently their surface have been coated by ethylene diamine. This gives a stable carbon nanoparticle of ~100 nm diameter. The characteristic absorption band of the Ru doped amine coated carbon nanoparticle (Ru:CNDEDAs) with peak maximum at

350 nm was observed in a simple absorption spectrophotometer (Fig. 3). On the surface of those carbon nanoparticles sufficient number of carbonyl (C=O) and α,β unsaturated carbonyl (>C=C-C=O) groups have been present. Thus both π-π* and n-π* transitions have been possible. Fig. 3 shows a peak at 350 nm and a shoulder around 240 nm which are probably due to n-π* and π-π* transitions. UV/Vis/NIR absorption spectroscopy can also be used to determine the stability of a nanoparticle sample. As the particle gets destabilised the original absorption peak would decrease in intensity due to depletion of the stable particle and often the original peak would broaden or a new peak would form at a longer wavelength. S. Mandal et al. [5] investigated the stability of cystein capped silver nanoparticle with time and found that the characterized characteristic peak gets broadened and decreases in intensity.

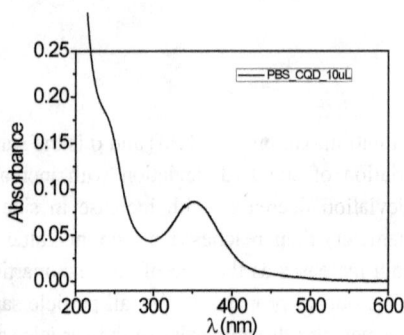

Figure 3. Absorption spectrum of Ru:CNDEDAs of ~100 nm diameter taken in a UV-Vis Jasco-650 spectrophotometer. Ru:CNDEDAs becomes the suitable replacement for standard dye molecule with high photo-stability and high quantum yield.

D. Paramelle et al. [6] established a relation between the size of Ag-nanoparticles with observed change in absorbance value. The observed extinction coefficients of light results from a combined effect of scattering and absorbance of light and illustrate the effect of interaction between matter and light. Thus, for a population consisting of equal concentration *but* of two different sizes of silver nanoparticles, the larger volume nanoparticles contribute more to the spectra or extinction coefficient value in comparison to the smaller one. In addition, as the extinction coefficient is being influenced by the size, the contribution of the scattering to the extinction coefficient increases for larger nanoparticle which results in broadening of the absorption spectrum when observed from bigger cluster of nanoparticles because of agglomeration [6].

3. Steady state fluorescence spectroscopy and imaging technique

In fluorescence spectroscopy, the molecular species/nanoparticles are first excited by absorbing a suitable fraction of light quanta from its ground state to one of the various vibrational levels of the excited electronic states. The phenomenon of absorption occurs within a time scale of $\sim 10^{-15}$ sec. Since the mass of electron is much less than the mass of the corresponding nuclei, a good approximation is that the electron cannot recognize the change in position or momentum of nuclei within the time scale of absorbance, then it attains relaxation to a new equilibrium geometry subsequent to the absorption process (Frank Condon approximation [7]). Collision with other molecules causes the excited species to lose the vibrational energy until it reaches back to the lowest vibrational level of the first excited electronic state. The molecules are further de-excited after coming down to any of the vibrational levels of the ground electronic state and thus emit photons in such processes. As the molecules drop down to any of the several vibration levels of the ground electronic state, the emitted photons exhibit different energies and thus possess different frequencies. Therefore, by analyzing the different frequencies of light emitted along with their relative intensities, the structure of different excited states can be determined. The various photo physical processes that happen after absorbing photons can be described through the well-known Jablonski diagram.

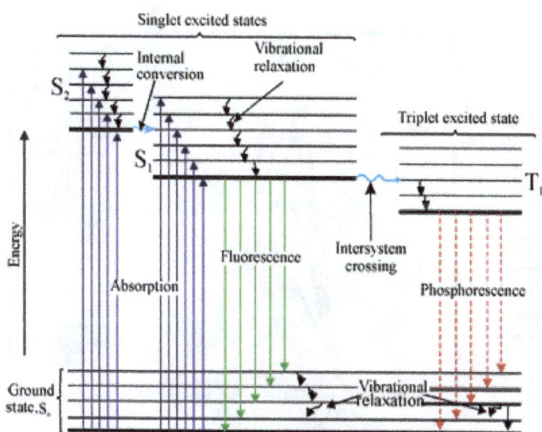

Figure 4. Typical Jablonski diagram indicating various photophysical processes [10].

In a typical fluorescence measurement, the excitation wavelength is fixed and the emission spectrum is recorded by monitoring emission intensity versus wavelength (λ), while in molecular absorption process the total range of white light frequencies are shown on the molecular species, resulting in the absorption of some particular frequencies (or wavelength), that yields the absorption spectrum.

Moreover, fluorescence spectroscopic technique can be used as an analytical tool for exploring the excited state phenomena of a special form of nanoparticle namely: quantum dots (explain briefly) [8] that have unique properties of fluorescence. Each sample has a particular wavelength where emission intensity has its maximum value called (λ_{em}^{max}) upon being excited at the absorption maxima/frequencies. In case of metal nanoparticles, smaller size nanoparticles have much stronger fluorescence than those of larger plasmonic nanoparticles. Fluorescence is thought to originate either from the quantum states within the metal core or the mixed ligand state at the organic inorganic interface i.e. organic ligands those are used as surface protecting agent [9]. However in the case of non metallic nanoparticles several school of thoughts have been proposed based on the quantum confinement effect. In a recent report Bera et al. proposed that florescence property developed in carbon nanoparticles synthesized during pyrolysis of citric acid and gradual improvement of aromatic conjugation between various aromatic moiety helps to establish emissive states [4].

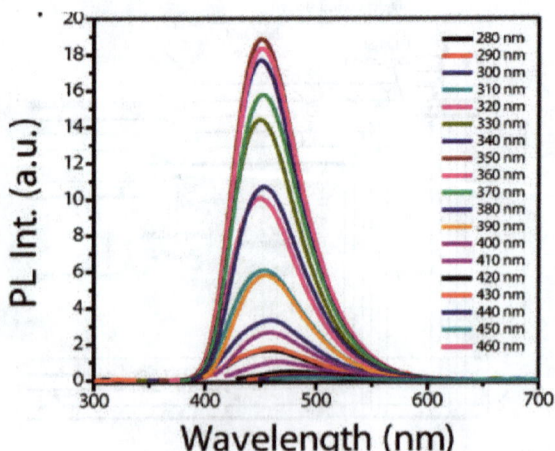

Figure 5. Fluorescence spectra of Ru(III)doped amine coated carbon nanoparticle at different excitation wavelength from 280 nm to 460 nm with 10 nm interval taken in Fluoromax-3 spectro fluorimeter.

Here, Fig. 5 represents the excited state spectral behavior of Ru (III) doped amine coated carbon nano particle (Ru:CNDEDAs) with various excitation wavelength starting from 280 nm to 460 nm with 10 nm interval. An emission maximum at 450 nm is observed at different excitation wavelength. The independence of fluorescence emission maxima on excitation wavelength for carbon nanoparticles provides a suitable replacement of standard fluorescence dye molecules.

Nanoparticles are used in various aspects in the field of bio-imaging, biomedical application or in energy storage purpose. In all those different application purposes the effectiveness of interaction pattern can be monitored primarily by observing the change in emission intensity of nanoparticle when they interact with those systems as a consequence of sensing the surrounding environment.

The technique usually applied in terms of either enhancement or quenching of fluorescence intensity refers to a process that results in either increase or decreases of fluorescence intensity repectively. A variety of processes can result in fluorescence quenching, such as excited state reactions, energy transfer, proton transfer, complex formation, collisional quenching, etc. [10].

Photoinduced electron transfer (PET) is the basic photophysical phenomenon that occurs in the photosynthesis process. The process of PET can be monitored by observing the changes in fluorescence intensity of donor (D) or acceptor (A) in presence of either A or D. Nanoparticles can act either as D or A in presence of each other. Sau et al. [11] investigated the phenomena of PET from Ru doped amine coated carbon nanoparticle (Ru:CNDEDAs) to a model electron acceptor menadione (MQ) which could be traced by observing the changes of fluorescence intensities of Ru:CNDED, with gradually increasing concentration of MQ. Fig. 4 depicts the changes in fluorescence intensities of Ru:CNDEDAs with increasing concentration of MQ from 1-10µM. In the excited state an electron is transferred from highest occupied molecular orbital (HOMO) of Ru:CNDEDAs to the lowest unoccupied molecular orbital (LUMO) of MQ. Thus excited population of Ru:CNDEDAs decreases, which is reflected from the decrease of emission intensity. The inset of Fig. 3 represents the nature of change of quenching pattern which can be quantified by observing the relative change of fluorescence intensity (F_0/F) against quencher concentration by famous Stern-Volmer equation [10].

$$F_0/F = 1 + K_{sv}[Q] \quad \ldots\ldots\ldots \tag{iii}$$

where K_{sv} is Stern-Volmer quenching constant and [Q] concentration of quencher. A linear change indicates the fact that only one type of quenching process is existent, either dynamic or static quenching, whereas non-linearity represents contribution from both dynamic and static quenching. The type of quenching i.e. either static or dynamic can be understood clearly with the help of time resolved techniques, where fluorescent lifetime remains invariant for static quenching but varies with the increase in the concentration of the quencher in dynamic quenching.

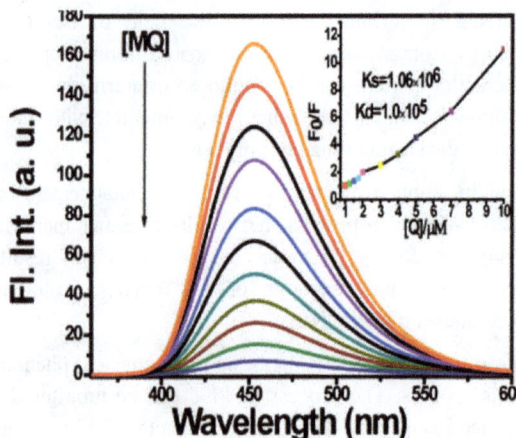

Figure 6. *Steady state fluorescence spectra of Ru:CNDEDAs with increasing concentration of quencher MQ from 1-10µM in water upon excitation at 350 nm taken in Fluoromax-3 spectro fluorimeter. Inset represents corresponding Stern-Volmer plot.*

3.1 Fluorescence imaging technique

Synthetic nanoparticles are emerging as versatile tool for biomedical applications, particularly in imaging. Nanoparticles with diameter 1-100 nm have dimensions comparable to biological functional units. Diverse surface chemistries, unique magnetic properties, tunable absorption and emission properties and recent developments in the site specific binding of nanoparticles suggests their potential as a probe for detection of diseases such as cancer. Aiming to the fact of delivery to a targeted site, the size, shape and hydrophilicity of the particles become pertinent important properties. Polyethylene glycol coated nanoparticles have been emerged as imposed hydrophilicity, which helps non specific adsorption of serum protein *in vivo*, thus providing longer circulation times [12]. Al-Jamal WT et al. explained their study on positively charged nanoparticles that have been designed to enhance cell penetration or phagocytosis for cell labeling [13]. Apart from these, the non-fluorescent tissue or biological samples can be visualized by staining the particular tissue or body part with fluorescent nanoparticles and subsequently monitoring the fluorescence image of that nanoparticle with different excitation sources with time. Confocal microscopic technique [14] is a suitable tool by which one can monitor in real time the relative change in structure of any biological tissue or species or we can also monitor the cytotoxic effects of a particular nanomaterial which has been synthesized with special engineering technique, keeping in mind its specific function and specificity towards a particular site. Ru doped amine coated carbon nanoparticles have been synthesized and modified with homocystein thiolactone by Bera et al. [4], that provides enhanced chemical reactivity and optimized distance for photoluminescence quenching by directly reacting with carcinogenic quinone inside the cell [4]. Fig. 7. describes the photo stability of Ru:CNDEDAHCTLs inside HeLa cell and also the sensing efficiency of carcinogenic quinone via depletion of fluorescence intensity by electron transfer from Ru:CNDEDAHCTLs to quinone [4].

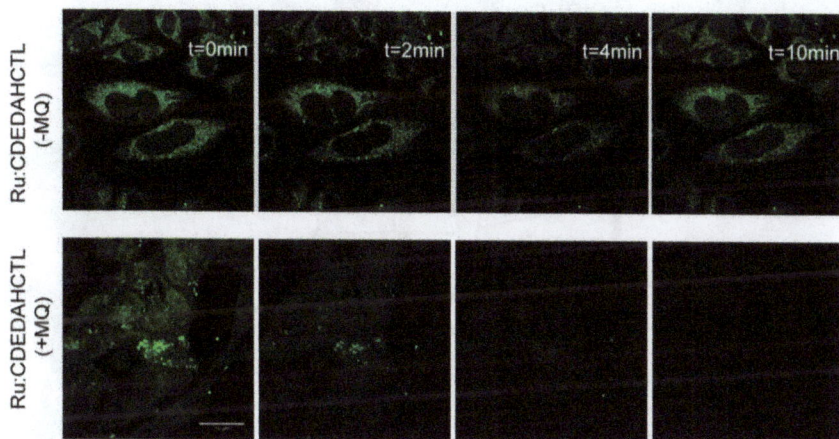

Figure 7. Fluorescence microscope image of Ru:CNDEDAHCTLs inside HeLa cell at different time point upon exposure to laser irradiation of 405 nm, shows high photo-stability of Ru:CNDEDAHCTLs (top panel) and bottom panel shows the decrease of fluorescence intensity of Ru:CNDEDAHCTLs inside HeLa cell upon treatment with model electron acceptor MQ with time.

From the bottom panel of above Fig. 7. it is clear that electron transfer takes place from Ru:CNDEDAHCTLs to menadione (MQ), a model quinine drug, that is why the fluorescence intensity decreases with time after treatment with MQ. In a confocal fluorescence microscope [14] (Fig. 8), the sample or specimen is generally excited by a laser light source. Light generated from the laser source passes through an (excitation) pinhole, is reflected by a dichroic mirror, and focused by a microscope objective to a small spot in the specimen. The dichroic mirror selectively transmits fluorescence light within longer wavelength region, while the excitation light with shorter wavelength cannot go through it. Specific dichroic mirrors can be made for the relevant wavelength regions of excitation and fluorescence.

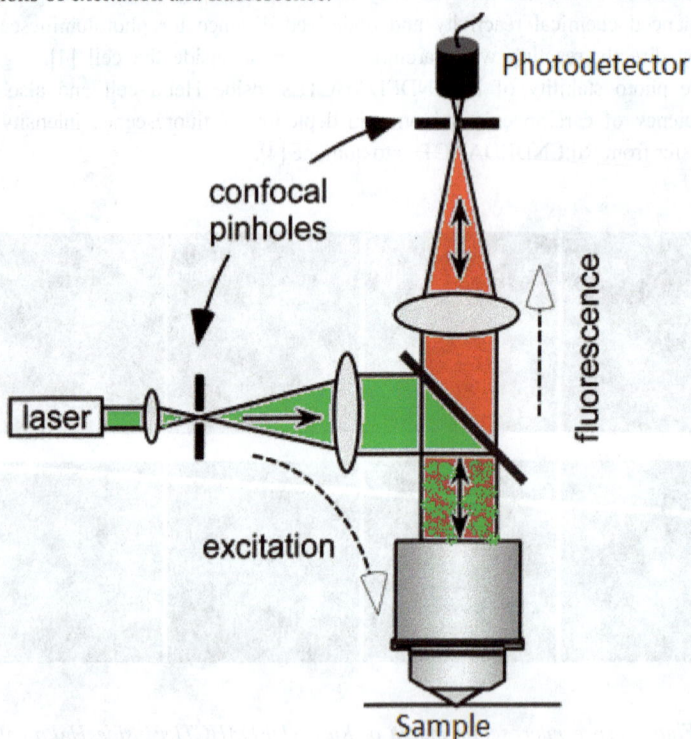

Figure 8. Schematic diagram of the principle of confocal fluorescence microscopy. Green and red colors are used for denoting excitation and fluorescence light [15].

In the above Fig. 8 the green light from laser source passes through a confocal pin hole, and falls upon a collimating lens and gets reflected by a dichroic mirror and finally hits upon the sample or specimen. Generated fluorescence light travels in the same path passing through the dichroic mirror and upon getting focused by a focusing mirror is finally fed to the detector.

3.2 Time resolved fluorescence spectroscopic technique

The time resolved technique is growing as emerging field in fluorescence spectroscopy particularly when one studies the fluorescence as a dynamic process. Moreover, time resolved measurement contains more information than is available from steady state data, as it has been already mentioned in the previous paragraph that time resolved measurement can be used to distinguish between static quenching and dynamic quenching. In static quenching a ground state complex formation takes place which does not decrease the decay time of the uncomplexed fluorophore because only the life time of uncomplexed fluorophores is estimated in time resolved measurement. During dynamic quenching however, an excited fluorophore undergoes a collision with a quencher molecule or other molecules which remain in the ground state, thus excited molecule undergo rapid deexcitation to the ground state and eventually lifetime of the excited fluorophore decreases.

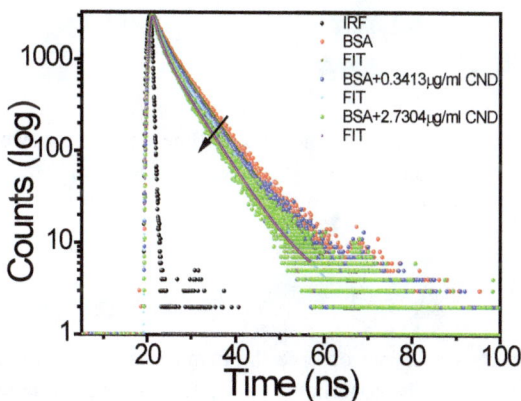

Figure 9.   Fluorescence decay traces of BSA protein (red) and with increasing concentration of Ru:CNDEDAs (blue and green respectively) in PBS buffer solution [18]. The emission wavelength has been monitored at 350 nm upon excitation at 295 nm from a nano-LED source. FWHM of prompt for excitation source 295 nm nano LED has been~980ps(black: scatter dots).

Time domain and frequency domain data acquisition methods are commonly used to determine the fluorescence lifetime of fluorophores. However, the instrumentation and data acquisition methods for each technique are different, both approaches are mathematically equivalent and their data can be inter-converted through Fourier transform. In time domain measurement the sample is excited with a short pulse of light (pulse width< 1-2 ns) from flash lamp, laser diode, pulsed laser or LEDs with sufficient delay between pulses. Now-a-days a variety of lifetime measurement techniques are available but the advantage of time correlated single photon counting technique (TCSPC) [16,17] has simplified data collection and enhanced quantitative photon counting. Photomultiplier tube or avalanche photodiode are used to record the time dependent distribution of emitted photon after each pulse.

In this context, Fig. 9 represents the emission decay traces of BSA protein of 2 μM concentration monitored at 350 nm, excited by 295 nm nano LED source measured using a TCSPC instrument from Horiba Jobin Evon. The emission decay trace of BSA protein is shown in red dots, the decay kinetics becomes faster with increasing concentration of Ru:CNDEDAs at two different concentration at 0.3413μg/ml and 2.7304 μg/ml shown in blue and light green scattered points. In TCSPC technique a fluorophore is excited by a sharp pulse of light photons resulting in the initial population $n_{(0)}$ of the fluorophore in the excites state. The excited state population decays with a rate $(\Gamma+K_{nr})$ following the equation:

$$\frac{dn(t)}{dt} = (\Gamma + K_{nr})n(t) \quad ......$$ (iv)

where $n_{(t)}$ is the number of excited molecules at time t following excitation, τ is the emissive rate, and k_{nr} is the nonradiative decay rate. This results in an exponential decay of the excited state population,

$$n_{(t)}= n_{(0)}exp(-t/\tau) ....$$ (v)

In a typical experiment we do not observe the number of excited molecule but rather fluorescence intensity, which is proportional to $n_{(t)}$. Thus number of molecules should be replaced by time dependent intensity which follows the usual expression for a single exponential decay. Thus,

$$I_{(t)} = I_0 \exp(-t/\tau)........................$$ (vi)

The fluorescence lifetime can be determined from the slope of a plot of log $I_{(t)}$ versus t, but more commonly by fitting the data to assumed decay models.[10]

In the above example the emission decay trace of BSA protein is fitted with a bi-exponential model following the equation.

$$I(t) = \sum_i \alpha_i \exp(-t/\tau_i) \quad \text{...} \quad \text{(vii)}$$

where $\sum_i \alpha_i$ normalised to unity.

In this particular example an instrument response function (prompt) is obtained by reflecting the photons from the excitation pulse through a non fluorescence scattering medium. In presence of fluorophore the slope of the decay curve is less steep because of the presence of finite population of the excited state. The decay kinetic of BSA protein is fitted with a bi-exponential fitting parameter resulting two lifetime values of 4.2 ns with (38%) relative contribution and 7.07 ns with (62%) contribution depicted in Table-1.

With increasing concentration of Ru:CNDEDAs the lifetime values of both the components decrease but the lifetime of the faster decaying component decreases rapidly in comparison to the slower one. Actually upon interaction with Ru:CNDEDAs, BSA protein undergoes some structural changes which might help to change in the environment of two tryptophan moieties.

Table-1   Estimated time constants with relative amplitude of BSA protein in absence and in presence of different concentration of carbon nanoparticles.

System	τ_1 (ns)	a_1	τ_2 (ns)	a_2	$<\tau>$ (ns)	χ^2
BSA	4.27	38%	7.17	62%	6.1	1.001
BSA+CND 0.3413 µg/ml	2.53	16%	6.68	84%	5.95	1.104
BSA+CND 2.7304 µg/ml	1.67	12%	6.32	88%	5.76	1.24

Where τ values indicates the lifetime values, a denotes the relative contribution of respective lifetime values and χ^2 value defines the weighted sum of the square of the deviation of experimental response to that of the calculated ones and in least square method the value of χ^2 should be close to 1 for a good fit. [19]

Thus exposed tryptophan becomes more accessible to Ru:CNDEDAs as confirmed from the change in the lifetime values from 4.27 ns to 1.67 ns at highest concentration of Ru:CNDEDAs, whereas other tryptophan moiety is little disturbed as it is less accessible by foreign particle as confirmed from the change in lifetime value from 7.17ns to 6.32ns at highest concentration of Ru:CNDEDAs. Bera et al. [4] synthesized Ru(III) doped amine coated carbon dots and another variant of it using the same Ru(III) as dopant but without amine coating. The size of Ru(III) doped CDs has been ~20 nm and was surface coated with amine using ethylene di-amine, but then the nanoparticle size enhances to ~80nm.

Carbon dots behaves as fluorophore because of the presence of a lot of $>C=C<$ and α, β unsaturated $>C=C-C=O$ on the surface of it. When carbon dots are coated using ethylene diamine, it binds with C=C bond through Michael Addition reaction thus providing not only as coating agent but at the same time convert some C=C bond to single bond which might help to reduce the loss of excitation energy through energy dissipation via cross linked C=C bond. Thus its fluorescence intensity increases and decay kinetics also follow the same trend as shown in Fig. 10 where blue dots and red dots represents the decay kinetics of Ru(III)CDs and amine coated Ru(III)CDs. From this figure it is clear that slope of the decay profile of Ru(III)CDs is much steeper in comparison to the amine coated Ru(III)CDs which signifies that without surface coating carbon dots have lifetime much less than carbon dots with amine coated Ru(III)CDs.

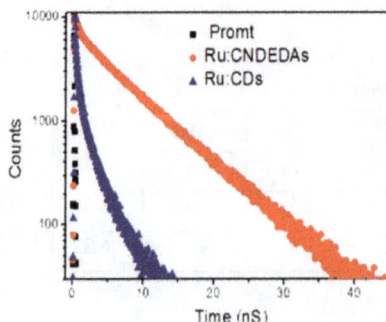

Figure 10. Fluorescence decay traces of Ru:CD (blue) and Ru:CNDEDA (red) in water excited at 375 nm and emission monitored at 450 nm, measured with TCSPC technique [4].

The lifetime of Ru doped carbon dots has small value in comparison to the Ru doped amine coated carbon nanoparticle. The lifetime enhances from 5.8 ns to 10.36 ns after surface coating with ethylene diamine of Ru(III)CDs when excited the sample at 375 nm sources. The enhanced lifetime of Ru(III) carbon dots are due to the protection of excitons from being exposed to the photoluminescence quenching sites in comparison to the carbon dots without any doping [4].

4.   Femtosecond fluorescence upconversion studies

The rapid development of biomedical research particularly in drug delivery, diagnostics of infectious diseases, methodologies for forensic application, etc. posses a great challenge to chemists, to find more efficient biological labels than traditional dye molecules which are resistant to photobleaching, non toxic, biocompatible, monochromatic highly luminescent and finally most importantly ultrasensitive in both *in vitro* and *in vivo* bioassay. Various semiconductor quantum dots (QDs) with size tunable luminescence property, high quantum yield, broad excitation spectrum and narrow emission spectrum can be successfully applied in biological analysis [20-26]. These QDs work well in laboratory condition where interfering process are comparatively few, however, when these QDs are being applied to *in vivo* condition, a significant problem arises because of the presence of background signal coming from interfering fluorescent bio-molecules, green fluorescence protein, etc. present in biological environment which are also excited simultaneously by UV radiation reducing the sensitivity of detection. The sensitivity can be improved by taking the advantage of using Förster's Resonance Energy Transfer (FRET) or Photo Induced Electron Transfer (PET) properties. In case of FRET the nanoparticles acting as either Donor (D) or Acceptor (A) with respect to other biomolecules can participate in radiative energy transfer if an overlap region is found between donor emission and acceptor absorption. Photo induced electron transfer property can also be used as a suitable ruler, for that, however, the surface of nanoparticle needs to be modified accordingly. Dongho Kim et al. studied emission decay dynamics on gold nanoparticles of 25 nm diameter that might depend on electron-hole recombination process using the femto-second up conversion technique. In their study, they used different excitation energy to generate thermal electrons and relaxation dynamics of those thermal electrons closely follow the emission decay dynamics of bare gold nanoparticles and concluded that light emission from gold nanoparticles are generated from surface plasmons [27]. In a typical upconversion experiment the exciting pulse may be the fundamental wavelength or its second or third harmonics, for example the FOG100, a fluorescence upconversion setup coupled with a MaiTai oscillator, generates 800 nm ~80 fs fundamental light. Doubling or tripling of frequencies are easily obtained by passing the fundamental wavelength through a nonlinear crystal like beta Barium Borate (BBO) crystal as used in FOG100 model. After passing through the BBO crystal doubling or tripling of frequency i.e. half and one third wavelength of fundamental wavelength are generated along with unconverted fundamental wavelength. A dichroic mirror is used which transmits the leftover fundamental wavelength and reflects the generated frequencies which then redirect to excite the fluorescence sample, which is termed as 'pump pulse'. Only the fluorescence signal is collected by using a cut off filter which efficiently filters out the unconverted excitation wavelength and passes only fluorescence light. A part of the fundamental light usually termed as 'probe' light transmitted through the dichroic mirror now travels through an optical delay line and meets with fluorescence light after certain time interval in femto-second time scale and is used to obtain information in the excited state. The gate pulse and fluorescence light are being collimated and converged using a focusing lens. Finally both the lights overlap at a point on a second BBO crystal which generates an upconverted signal by maintaining the correct phase matching condition with respect to the both lights. The fundamental light which is used to generate the upconverted signal in a BBO crystal with fluorescence signal is called optical gate pulse because it controls the generation of upconverted photon, in the absence of which, no upconverted photon could be detected by a photo multiplier tube (PMT) detector and thus it acts as an optical gate. The upconverted signals are allowed to pass through another filter which allows only upconverted photons and discard fluorescence and fundamental signals. Gate pulse allows simultaneous delay with respect to the zero timing of excitation of the sample. The photons in all the delay timing starting from zero to some higher timing in pico-second time domain and a histogram of intensity against time is constructed. Fitting the exponential decay profile of fluorescence materials, decay constants are obtained which give information in fast time scale for rapid phenomena like proton transfer, PET, FRET, etc.

For an example the electron transfer from Tryptophan moiety of HSA to carbon nanoparticle has been characterized through femtosecond fluorescence upconversion technique. The decay profiles of Ru:CNDEDAs in absence and with increasing concentration of HSA are shown in Fig. 12.

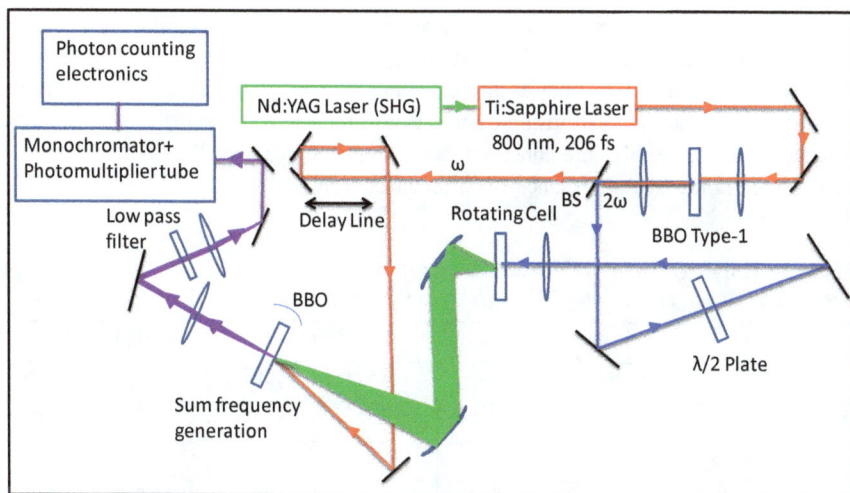

Figure 11. Schematic diagram of Femtosecond fluorescence upconversion setup of FOG100 in Saha Institute of Nuclear Physics- Kolkata.

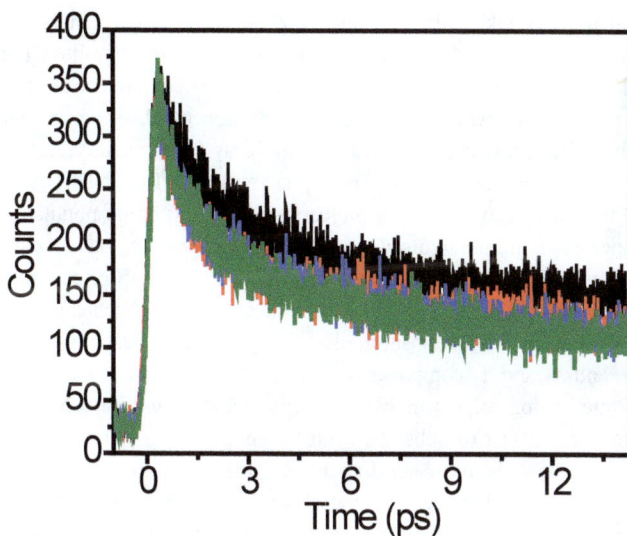

Figure 12. Representative femtosecond resolved fluorescence transient of Ru:CNDEDAs in absence (black)and in presence of 115μM (red), 231μM (blue) and 478μM (green) HSA respectively the monitored at 450 nm in PBS buffer.

Ru:CNDEDAs has been excited with a frequency doubled 370 nm exciting pulse and the upconverted photons have been collected simply by mixing gate pulse of 740 nm with fluorescence maxima at 450 nm at different time intervals. All data are fitted with the help of the igor pro software using a three exponential function and fitted parameters. Change of the ultrashort time constants considering the decay of Ru:CNDEDAs up to 15 ps time regime in the presence of different concentration of HSA are furnished in Table 2.

Table 2. Estimated ultrafast, fast and longer lifetime of Ru:CNDEDAs in absence and in presence of donor 116µM and 231µM HSA in PBS buffer solution.

System	τ_1 (fs)	a_1	τ_2 (ps)	a_2	τ_3 (ps)	a_3
CNDs	1499	34%	5.671	30%	254	36%
CND+HSA 116µM	1055	53%	7.553	11%	48.554	36%
CNDs+HSA 231µM	927	46%	5.957	25%	47.94	29%
CNDs+HSA 478µM	1145	49%	7.012	26%	67.538	25%

Actually electron transfer takes place in such a time scale which can be monitored accurately in picoseconds to sub-picosecond time scale. From the fitted parameters obtained through decay profiles, it is clear that the ultrashort component i.e. τ_1 and long component i.e. τ_3 decreases with increasing concentration of HSA protein whereas a change in τ_2 is insignificant. Continuous decrease of decay components clearly indicates the phenomenon of electron transfer from Tryptophan moiety to Ru:CNDEDAs. Vauthey et al. postulated that in a redox system a molecule always has some populations of its photo excited species at minimal distance where charge transfer can occur without significant diffusion [28]. Diffusion has no contribution for this part of electron transfer and occurs within a very short time scale as reflected through change of the (τ_1) component. According to this proposition the change in τ_1 is responsible for direct electron transfer and τ_2 and τ_3 components describe the diffusion controlled electron transfer through intervening solvent molecules. Sengupta et al. used a series of acridine derivatives to study the effect of substituent on the intercalating efficiency of acridine derivative in comparison to the basic acridine moiety with DNA molecules. Intercalating efficiency was probed through photo induced electron transfer phenomena from DNA base pair to acridine derivative in femtosecond time regime using fluorescence up-conversion technique. They conclude that electron transfer not only depends on drug molecular structure but on the water structure in the vicinity of DNA molecules. Thus their study helped in future drug design and delivery system [29]. In another study

Sengupta et al. shaded lights on electron transfer property between an electron acceptor Proflavin with two donor like triethyl amine and dimethyl aniline and pair in medium of different polarity and reported formation of various transient species such as contact ion pair (CIP), solvent separated ion pair (SSIP). They also reported that the fastest electron transfer and solvation in early time scale could be modulated in the heterogeneous medium which is not possible in a homogeneous medium [30].

5.   Laser flash photolysis

In the emerging field of nano technology, particularly in nanomedicine, nanoparticles are frequently used because of their compatibility, relatively less toxicity, preferential surface modification and are easily eliminated from the body after the performance of their precise functions. From the above mentioned techniques like steady state UV-Vis absorption, fluorescence, time resolved fluorescence and ultrafast time resolved fluorescence we can get useful information regarding ground state as well as excited state phenomena. Nanoparticles can act as electron donor as well as electron acceptor in the presence of suitable electron donor or acceptor in close proximity of the nanoparticle. Thus photo induced electron transfer (PET) or energy transfer (FRET) can happen to and from nanoparticles within a close microenvironment. Particularly, PET is the most commonly occurring phenomenon in drug-nanoparticle and protein-nanoparticle interactions which can be best studied using laser flash photolysis coupled with external magnetic field [26]. Laser flash photolysis coupled with an external magnetic field not only detects the occurrences of PET but also authenticates the form of the initial spin state of the radicals/radicals ions are in.

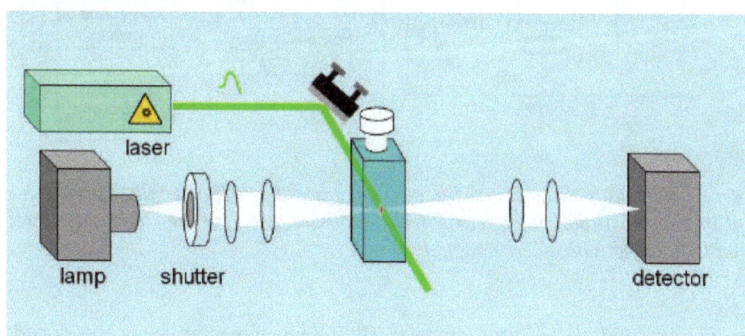

Figure 13.   Schematic representation of principle of laser flash photolysis.

Page No-96:The magnetic field effect on the newly generated transient species i.e. RPs/RIPs can be carried out by passing direct current through a pair of electromagnetic coil placed inside the sample chamber and the strength of the magnetic field can be varied from 0.0 to 0.08 Tesla [31,32].

In a typical nanosecond laser flash photolysis set up, the sample is excited by a Nd:YAG laser with its third harmonics (355 nm) 10 mJ energy with full width at half maxima (FWHM) ≈ 8 ns. Absorption of light from a pulse Xenon Lamp (150 Watt) is used to detect newly generated transient species in the system. The photomultiplier output has been fed into a Agillent Infiniium oscilloscope (DSO 8064A, 600 MHz, 4Gs/s and data have been transferred to the computer through IYONIX software. The magnetic field effect on the newly generated transient species i.e. RPs/RIPs can be carried out by passing direct current through a pair of electromagnetic coil placed inside the sample chamber and the strength of the magnetic field can be varied from 0.0 to 0.08 T [31,32].

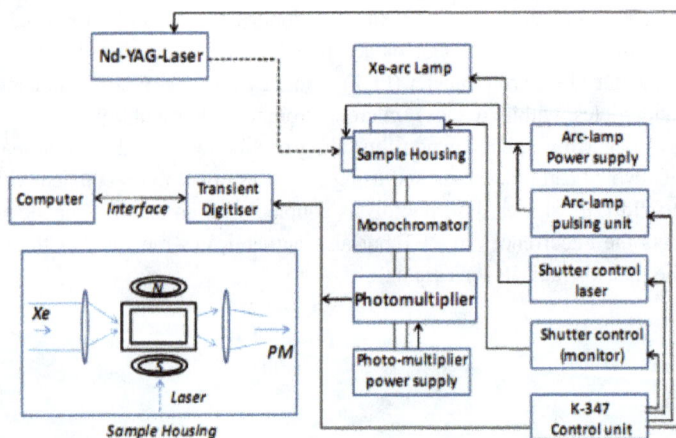

Figure 14. Schematic diagram of different component of the laser flash photolysis set-up present in Saha Institute of Nuclear Physics-Kolkata, used for the said study. Inset shows the sample housing with electromagnet.

Sau et al. reported an electron transfer (PET) phenomenon from Ru doped amine functionalised carbon dots (Ru:CNDEDAs) to 2-Methyl-1,4-naphthoquinone or commonly known as Menadione (MQ) [11] for the first time. In the laser flash photolysis experiment Ru:CNDEDAs is excited by the third harmonics of Nd:YAG laser. The excited singlet Ru:CNDEDAs ([1]Ru:CNDEDAs*) either comes down to the ground state or undergo intersystem crossing (ISC) to generate corresponding triplet state ([3]Ru:CNDEDAs). The nonfluorescent transient triplet species are quantified by monitoring absorbance at different wavelength through their absorbance to higher triplet state by absorbing pulsed Xe light. The absorption spectrum of Ru:CNDEDAs shows a band at 470nm with a shoulder at 440 nm whereas that of MQ shows a absorption peak at 360nm (Fig.15).

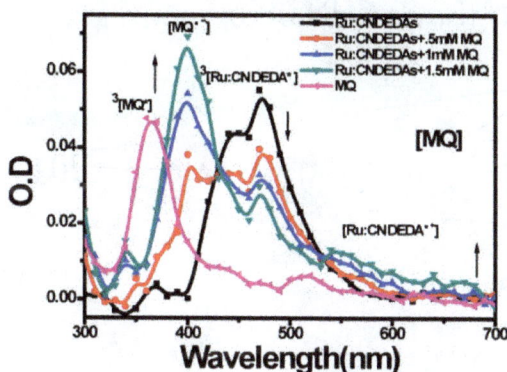

Figure 15. Transient triplet–triplet absorption spectra of Ru:CNDEDAs,1 μs after laser flash in phosphate buffer (pH= 7.4) with increasing concentrations of MQ (0.5–1.5 mM) (λ_{ex} = 355 nm).

With increasing concentration of MQ the absorption band at 470 and 440 nm decreases with the concomitant enhancement of a new peak at 400 nm and a broad band from 550-600 nm through two isosbestic points at 430 nm and 530 nm which signify that new species are generated at the expense of triplet [3]Ru:CNDEDAs in presence of MQ. Sau et al. reported earlier that the peak at 400 nm occurs because of a [MQ·[-]] radical anion. Therefore, it is quite reasonable to understand that PET takes place from [3]Ru:CNDEDAs to MQ with the formation of [MQ·[-]] and corresponding [Ru:CNDEDAs·[+]] which shows absorbance around 550-600 nm [11].

Magnetic field effect is an efficient tool to identify the geminate radical ion pair or radical pair which generate due to the photo induced electron transfer, with their initial spin state even in the presence of other neutral or non geminate radical ion pair [33-40]. After photo excitation by a laser flash a geminate radical ion pair/radical pair forms between donor and acceptor as transient intermediate might undergo maximum intersystem crossing between degenerate $S_0 \leftrightarrow T_0$, $T_{\pm}$ through spin flipping in presence of an internal magnetic field i.e. hyperfine interaction present within the system when the spin correlated radical ion pairs are separated by a certain distance where exchange interaction becomes negligible (Scheme-2).

(a)

$$D \longrightarrow {}^1D^* \longrightarrow {}^3D$$

$${}^3D + {}^1A \longrightarrow {}^3D^+ \, {}^{\cdot}A^-$$

$${}^3(D^+\uparrow......\uparrow\!\dot{A}^-) \xrightarrow{\text{ISC}} {}^1(D^+\uparrow......\downarrow\!\dot{A}^-)$$

(b)

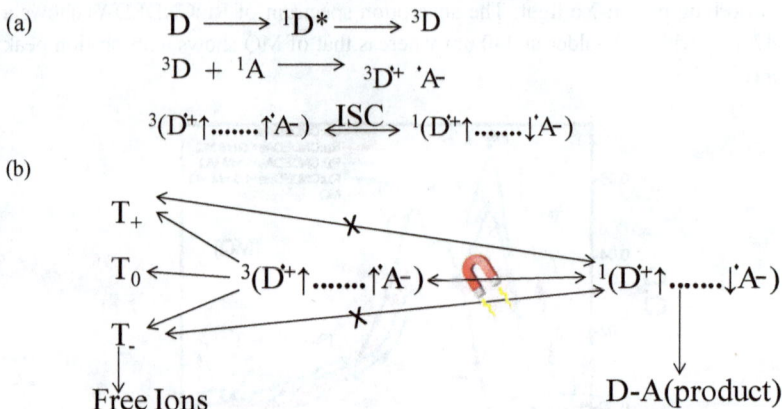

Figure 16. (a) Schematic representation of donor ($^3D^*$) and acceptor (A^-) radical generation (black vertical arrow represents the spin correlation between the geminate free electron either triplet (3) or singlet (1) and (b) Zeeman splitting of degenerate triplet state in presence of magnetic field which leads to either free ion formation or recombination products.

Now if an external magnetic field is applied to the system which is sufficient to overcome the hyperfine interaction that reduces the chance of intersystem crossing via Zeeman splitting (Scheme-2b) of the degenerate triplet state and induces the formation of recombination product or free ion formation which is the signature of the initial spin state of the radical ion pairs either singlet or triplet respectively [41].

5.   Conclusion

The natural processes of light absorption, reflection, transmission and emission have been very intimately related to life science since the inception of the biological world on earth and our ancestors understood the science through basic tools and genuine perceptions. Understanding of the photochemical processes today has evolved advanced experimental techniques. This article has paved a simple way to explain the phenomenology of complicated interactions between nano-particle and biological systems. Apart from the steady state systems, the kinetic systems depict the state-of-the-art of investigating the excited state processes down to a scale of sub-pico or femto seconds which suit the miniature world of the nanomaterial forum, where molecular interaction is the cardinal issue. Imaging through fluorescence emission in confocal microscopy puts a step forward for investigating the cellular processes *in-vitro* in the laboratory. Probing the transient species through nanosecond laser flash photolysis is a step forward to gain insight into the photochemical processes. The effect of a magnetic field in the biosphere is not negligible, as we know that the geomagnetic field ($\sim$ 0.25 to 0.65 gauss) at the surface of earth could be responsible for various subtle bio-phenomenon switched in the presence of natural sunlight. Although, much has been discussed in this concise article, perhaps these are not all, but the future will invoke more questions and thus newer and emerging methods, would help expand the level of our understanding.

References

[1]   N. Kometani, M. Tsubonishi, T. Fujita, K. Asami, and Y. Yonezawa, Preparation and Optical Absorption Spectra of Dye-Coated Au, Ag, and Au/Ag Colloidal Nanoparticles in Aqueous Solutions and in Alternate Assemblies, Langmuir 17 (2001) 578-580. https://doi.org/10.1021/la0013190

[2]   S. Link and M.A. El-Sayed, Size and Temperature Dependence of the Plasmon Absorption of Colloidal Gold Nanoparticles, J. Phys. Chem. B 103(1999) 4212-4217. https://doi.org/10.1021/jp984796o

[3]   U. Kreibig, L. Genzel, Optical absorption of small metallic particles, Surf. Sci. 156 (1985) 678 -700. https://doi.org/10.1016/0039-6028(85)90239-0

[4]   K. Bera, A. Sau, P. Mondal, R. Mukherjee, D. Mookherjee, A. Metya, A.K. Kundu, D. Mandal, B. Satpati, O. Chakrabarti, and S. Basu, Metamorphosis of Ruthenium-Doped Carbon Dots: In Search of the Origin of Photoluminescence and Beyond, Chem. Mater. 28 (2016) 7404−7413. https://doi.org/10.1021/acs.chemmater.6b03008

[5]   S. Mandal, A. Gole , N. Lala , R. Gonnade , V. Ganvir and M. Sastry, Studies on the Reversible Aggregation of Cysteine-Capped Colloidal Silver Particles Interconnected via Hydrogen Bonds, Langmuir, 17 (2001) 6262-6268. https://doi.org/10.1021/la010536d

[6]   D. Paramelle, A. Sadovoy, S. Gorelik, P. Free, J. Hobleya and D.G. Fernig, A rapid method to estimate the concentration of citrate capped silver nanoparticles

from UV-visible light spectra, Analyst, 139 (2014) 4855–4861. https://doi.org/10.1039/C4AN00978A

[7]   W.L. Smith, Spectrochimica Acta Part A 72 (2009) 915- 1134. https://doi.org/10.1016/j.saa.2008.12.025

[8]   U.R. Genger, M. Grabolle, S.C. Jaricot, R. Nitschke, and T. Nann, Quantum dots versus organic dyes as fluorescent labels, Nature Methods, 5 (2008) 763 – 775. https://doi.org/10.1038/nmeth.1248

[9]   B.A. Ashenfelter, A. Desireddy, S.H. Yau, T. Goodson III, and T.P. Bigioni, Optical Properties and Structural Relationships of the Silver Nanoclusters $Ag_{32}(SG)_{19}$ and $Ag_{15}(SG)_{11}$, J. Phys. Chem. C, 121 (2017) 1349–1361. https://doi.org/10.1021/acs.jpcc.6b10434

[10] Joseph R. Lakowicz, Principles of Fluorescence Spectroscopy Third Edition, pages 278-327, Springer, 2006.

[11] A. Sau, K. Bera, P. Mondal, B. Satpati, and S. Basu, Distance-Dependent Electron Transfer in Chemically Engineered Carbon Dots, J. Phys. Chem. C, 120, (2016) 26630−26636. https://doi.org/10.1021/acs.jpcc.6b08146

[12] G. Prencipe, S.M. Tabakman, K. Welsher, Z. Liu, A.P. Goodwin, L. Zhang, J. Henry, and H. Dai, PEG branched polymer for functionalization of nanomaterials with ultralong blood circulation, J Am Chem Soc. 131 (2009) 4783–4787. https://doi.org/10.1021/ja809086q

[13] W.T. Al Jamal , K.T. Al-Jamal , P.H. .Bomans , P.M. Frederik , K. Kostarelos, Functionalised-quantum-dot-liposome hybrid as multimodal nanoparticles for cancer , Small, 4 (2008) 1406-15. https://doi.org/10.1002/smll.200701043

[14] S. Nie, D.T. Chiu, and R.N. Zare, Real-Time Detection of Single Molecules in Solution by Confocal Fluorescence Microscopy, Anal. Chem. 67 (1995) 2849-2857. https://doi.org/10.1021/ac00113a019

[15] Laser Scanning Confocal Microscopy, Nathan S. Claxton, Thomas J. Fellers, and Michael W. Davidson: https://www.aptechnologies.co.uk/images/Data/Vertilon/PP6207.pdf Department of Optical Microscopy and Digital Imaging, National High Magnetic Field Laboratory, The Florida State University, Tallahassee, Florida 32310.

[16] W. Becker, Advanced time-correlated single photon counting techniques. Springer; Berlin; New York: 2005. https://doi.org/10.1007/3-540-28882-1

[17] W. Becker, The bh TCSPC Handbook. Third. Becker & Hickl Gmbh; 2008.

[18] K. Malarkania, I. Sarkarb, and S. Selvama, In Press, Accepted Manuscript, https://doi.org/10.1016/j.jpha.2017.06.007.

[19] Bernard Valeur, Molecular Fluorescence Principles and Applications, ISBN 3-527-29919-X, pp179-184.

[20] W.C.W. Chan, S. Nie, Quantum dot bioconjugates for ultrasensitive nonisotopic detection, Science,281 (1998) 2016 – 2018. https://doi.org/10.1126/science.281.5385.2016

[21] M. Bruchez, D.J. Moronne, P. Gin, S. Weiss, A.P. Alivisatos, Semiconductor nanocrystals as fluorescent biological labels, Science 281 (1998) 2013 – 2016. https://doi.org/10.1126/science.281.5385.2013

[22] J.R. Taylor, M.M. Fang, S. Nie, Probing Specific Sequences on Single DNA Molecules with Bioconjugated Fluorescent Nanoparticles, Anal. Chem. 72 (2000) 1979 – 1986. https://doi.org/10.1021/ac9913311

[23] R.C. Bailey, J.M. Nam, C.A. Mirkin, J.T. Hupp, Real-Time Multicolor DNA Detection with Chemoresponsive Diffraction Gratings and Nanoparticle Probes, J. Am. Chem. Soc. 125 (2003) 13541 – 13547. https://doi.org/10.1021/ja035479k

[24] E.R. Goldman, E.D. Balighian, H. Mattoussi, M.K. Kuno, J.M. Mauro, P.T. Tran, G.P. Anderson, Avidin: A Natural Bridge for Quantum Dot-Antibody Conjugates, J. Am. Chem. Soc. 124(2002) 6378 – 6382. https://doi.org/10.1021/ja0125570

[25] S.J. Rosenthal, I. Tomlinson, E.M. Adkins, S. Schroeter, S. Adams, L. Swafford, J. McBride, Y. Wang, L.J. DeFelice, R.D. Blakely, Targeting Cell Surface Receptors with Ligand-Conjugated Nanocrystals, J. Am. Chem. Soc. 124 (2002) 4586 – 4594. https://doi.org/10.1021/ja003486s

[26] J.K. Jaiswal, H. Mattoussi, J.M. Mauro, S.M. Simon, Long-term multiple color imaging of live cells using quantum dot bioconjugates, Nat. Biotechnol. 21(2003) 47 – 51. https://doi.org/10.1038/nbt767

[27] Y.N. Hwang , D.H. Jeong , H.J. Shin , and D. Kim, Femtosecond Emission Studies on Gold Nanoparticles, J. Phys. Chem. B, 106 (2002) 7581–7584. https://doi.org/10.1021/jp020656+

[28] E. Vauthey, Investigations of bimolecular photoinduced electron transfer reactions in polar solvents using ultrafast spectroscopy, J. Photochem. Photobiol. A: Chem. 179 (2006) 1–12. https://doi.org/10.1016/j.jphotochem.2005.12.019

[29] C. Sengupta, and S. Basu, A spectroscopic study to decipher the mode of interaction of some common acridine derivatives with CT DNA within

nanosecond and femtosecond time domains, RSC Advances, 5 (2015) 78160-78171. https://doi.org/10.1039/C5RA13035B

[30] C. Sengupta, P .Mitra, B.K. Seth, D. Mandal and S. Basu, Electronic and spatial control over the formation of transient ion pairs during photoinduced electron transfer between proflavine–amine systems in a subpicosecond time regime, RSC Advances, 7 (2017) 15149 – 15157. https://doi.org/10.1039/C6RA28286E

[31] B.K. Seth and S. Basu, Research Methodology in Chemical sciences Experimental and Theoretical Approach, Chapter-1, pp 1-14. CRC Press, 2017.

[32] S. Aich and S. Basu, Laser flash photolysis studies and magnetic field effect on a new heteroexcimer between N-ethylcarbazole and 1,4-dicyanobenzene in homogeneous and heterogeneous media, J.Chem. Soc. Faraday Trans., 91 (1995) 1593-1600. https://doi.org/10.1039/ft9959101593

[33] D.R. Kattnig, E.W. Evans, V. Déjean, C.A. Dodson, M.I. Wallace, S.R. Mackenzie, C.R. Timmel, P. Hore, Chemical amplification of magnetic field effects relevant to avian magnetoreception, Nat. Chem. 8 (2016) 384–391.

[34] R. Nishikiori, S. Morimoto, Y. Fujiwara, A. Katsuki, R. Morgunov, Y. Tanimoto, Magnetic Field Effect on Chemical Wave Propagation from the Belousov–Zhabotinsky Reaction, J. Phys. Chem. A, 115 (2011) 4592–4597. https://doi.org/10.1021/jp200985j

[35] P.W. Atkins, T.P Lambert, Annu. Rep. Prog. Chem., Sect. A. Inorg. and Phys. Chem. 72 (1975) 67- 88. https://doi.org/10.1039/pr9757200067

[36] I.R. Gould, N.J. Turro, M.B. Zimmt, Magnetic Field and Magnetic Isotope Effects on the Products of Organic Reactions, Adv. Phys. Org. Chem. 20 (1984) 1-53. https://doi.org/10.1016/S0065-3160(08)60147-1

[37] U.E. Steiner, T. Ulrich, Magnetic field effects in chemical kinetics and related phenomena, Chem. Rev. 89 (1989) 51-147. https://doi.org/10.1021/cr00091a003

[38] K. Bhattacharyya, and M Chowdhury, Environmental and magnetic field effects on exciplex and twisted charge transfer emission, Chem. Rev. 93,(1993) 507-53. https://doi.org/10.1021/cr00017a022

[39] S. Aich, S. Basu, Magnetic Field Effect: A Tool for Identification of Spin State in a Photoinduced Electron-Transfer Reaction, J. Phys. Chem. A 102 (1998) 722-729. https://doi.org/10.1021/jp972264m

[40] T. Sengupta,. S.D. Choudhury and S. Basu, Medium-Dependent Electron and H Atom Transfer between 2'-Deoxyadenosine and Menadione: A Magnetic Field

Effect Study, J. Am. Chem. Soc. 126 (2004) 10589-10593.
https://doi.org/10.1021/ja0490976

[41]   D. Dey, A. Bose, M. Chakraborty, S. Basu, Magnetic Field Effect on
       Photoinduced Electron Transfer between Dibenzo[a,c]phenazine and Different
       Amines in Acetonitrile–Water Mixture, J. Phys. Chem. A 111 (2007) 878–884.
       https://doi.org/10.1021/jp0661802

Chapter 5

Characterization of Nanomaterials: X-ray Diffraction Method, Electron Microscopy and Light Scattering

Bichitra Nandi Ganguly

Saha Institute of Nuclear Physics, 1/AF Bidhannager, Kolkata, India

bichitra.ganguly@saha.ac.in

Abstract

This chapter reviews methods to illuminate the molecular architecture of nanoparticles through special high resolution techniques such as X-ray diffraction studies (XRD), which is used to examine the crystallinity of a sample, the powder diffraction method which assesses the size of the nano crystalline sample by Debye-Scherrer method. Transmission (TEM) and scanning electron microscopic (SEM) methods are used to get information on the structural morphology, elemental composition and size of the nanoparticles. Further, the dynamic light scattering (DLS) method gives an illustration of measurement of hydrodynamic radii of nanoparticles in its dispersed phase. These are some of the basic tools one needs to use for the physical characterization of nanoparticles. Some examples are also given in order to better understand the processes.

Keywords

X-Ray Diffraction, Debye Scherrer Method, Electron Microscopy, Light Scattering

Contents

1. Introduction 105

2. X-ray diffraction technique (XRD) 105

3. Electron microscopy 111

    2.1 Transmission electron microscope (TEM) 112

    2.2 Scanning electron microscope (SEM) 114

    2.3 Dynamic light scattering (DLS) 115

References 121

1. Introduction

Nanoparticles are of current interest because of their possible effects on human health and the environment and owing to the increasing output of various synthetic nanoparticles into the environment. Nanoparticles are used in many different applications and are created by many different processes. Their measurement and characterisation pose interesting analytical challenges.

Nanoparticles are in general entities with some billionths of a metre in size. The formal definition of a nanoparticle is a "nanoobject with all three external dimensions in the nano meter scale" although in practice the term is often used to refer to particles larger than 100 nm. The reason for this is that the behaviour of nanoparticles and the applicability of measurement techniques vary with size and environment, to the extent that 500 nm particles can either be considered very large or very small, depending on the frame of reference. These days, ultra small metallic nanoparticles are synthesized, which require sophisticated techniques to characterize them.

Characterisation techniques can be subdivided by both general measure and the phase in which the nanoparticles exist. Measurements of each type present their own difficulties and often have subtly different interpretations. Moreover, comparison of results between phases is very difficult, and matrix effects can be significant due to the high surface area to mass ratio of nanoparticles. The techniques presented below give a general overview of modern measurements made on nanoparticles for a range of applications.

Nanoparticles in the solid phase exist either as a powder or encapsulated in a solid medium. The former can take several forms including loose powders and wet or dry 'powder cakes' for convenience of handling. As such, any analysis must take into account how the particles will eventually be used because this will affect their final agglomeration state and other properties.

In this chapter, simple characterization methods used for the characterization of the size of nanoparticles either in solid powder form or in the liquid sol/dispersed form through several modern physical tools are discussed, viz: 1) X-ray diffraction technique, 2) Transmission and scanning electron microscopy, and 3) dynamic light scattering technique.

2. X-ray diffraction technique (XRD)

X-ray diffraction, a phenomenon in which the atoms of a crystal, by virtue of their uniform spacing, cause an interference pattern of the waves present in an incident beam of X rays. The atomic planes of the crystal act on the X-rays in exactly the same manner as does an uniformly ruled grating on a beam of light.

Bragg diffraction occurs when radiation, with a wavelength comparable to atomic spacing, is scattered in a specular fashion by the atoms of a crystalline system, and undergoes constructive interference. For a crystalline solid, the waves are scattered from lattice planes separated by the interplanar distance d. When the scattered waves interfere constructively, they remain in phase since the difference between the path lengths of the two waves is equal to an integer multiple of the wavelength. The path difference between two waves undergoing interference is given by $2d\sin\theta$, where θ is the scattering angle, see Fig. 1.

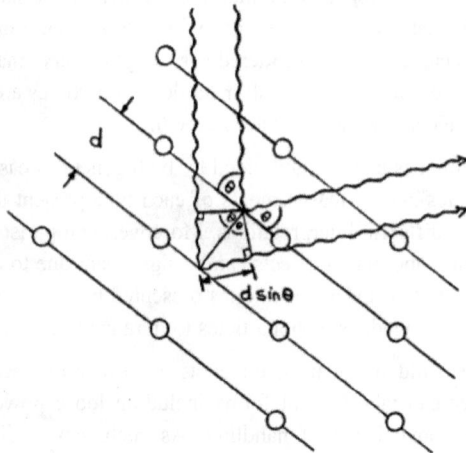

Figure 1. It can be seen that if the spacing between reflecting planes is d and the glancing angle of the incident X-ray beam is θ, the path difference for waves reflected by successive planes is 2d sin θ. Hence the condition for diffraction (the Bragg condition) is nλ = 2d sin θ where n is an integer and λ is the x-ray wavelength.

The equation for reflection (Bragg condition) can be satisfied for any set of planes whose spacing is greater than half the wavelength of the X-rays used (if $d < \lambda/2$, then $\sin\theta > 1$, which is impossible). This condition sets a limit on how many orders of diffracted waves can be obtained from a given crystal using a X-ray beam of a given wavelength. Since the crystal pattern repeats in three dimensions, forming a three-dimensional diffraction grating, three integers, denoted h, k, l are required to describe the order of the diffracted waves. These three integers, the Miller indices used in crystallography, denote the orientation of the reflecting sheets with respect to the unit cell and the path difference in units of wavelength between identical reflecting sheets. The wavelength path difference, m, between identical sheets is the greatest common divisor of h, k, l. The sheet orientation is the plane containing the three points found by moving along the OA axis= h/m units, along the OB axis= k/m units, and along the OC axis= l/m units, where one unit, a, is the length of the edge of the unit cell (lattice constant).

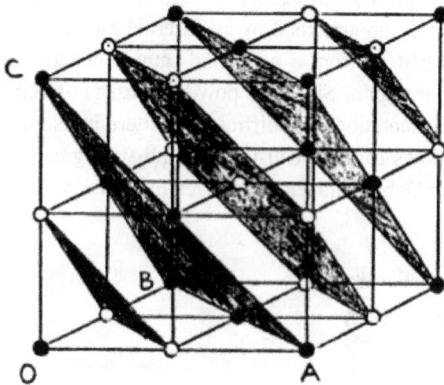

Figure 2. In this structural diagram (NaCl cubic crystal (111), the crystal axes are denoted by the letters OA, OB, OC), these being the intervals at which the crystal pattern repeats. Since NaCl is a cubic crystal, OA = OB = OC = a = the length of the edge of the unit cell, also called the lattice constant. In this structural diagram, the crystal axes are denoted by the letters OA, OB, OC, these being the intervals at which the crystal pattern repeats.

In the powder (**Debye-Scherrer**) method, the X-rays fall on a mass of tiny crystals in all orientations, and the diffracted beams of each order h, k, l form a cone. Arcs of the cones are intercepted by a film surrounding the specimen.

Let us consider first a single crystal. Consider one set of reflecting sheets of separation d_i. Only one angle, θ, exists for which reflection of the nth order occurs (Fig. 3a). Now the crystal is rotated about an axis along the incoming beam direction. The diffracted beam sweeps out a circle as shown in Fig. 3b. This assumes however, that the angle was chosen correctly to get a reflection. To ensure this is so, at least during part of the experiment, the crystal is rotated about an axis perpendicular to the incoming beam, (Fig. 3 c). Thus there will be a ring for each set of sheets in the crystal and each order of reflection from each set of sheets. Finally, there may have been sheets oriented parallel to the paper (Fig. 3d) and in order to get reflections from them, the crystal is rotated about a third axis perpendicular to the X-ray beam and parallel to the paper. The three rotations result in circular patterns for every sheet separation d_i and every order of reflection. Since the three rotations result in the crystal being oriented in every possible position in space, it is equivalent to use powdered crystals which, because of their random orientation, are already oriented in every possible direction. To further ensure randomisation, the powder sample mounting is rotated about axis 2 by an external motor. Thus, for monochromatic X-rays, the diffraction pattern from a powdered sample consists of one circle for each order of each sheet separation d_i. Since the powder consists of very small crystals of the material in all possible orientations, the diffraction pattern is a series of concentric circles. This type of pattern allows the unit cell to be found with great precision. This method was first used by Peter Debye and Paul Scherrer in 1916 and independently by Albert Hull in 1917.

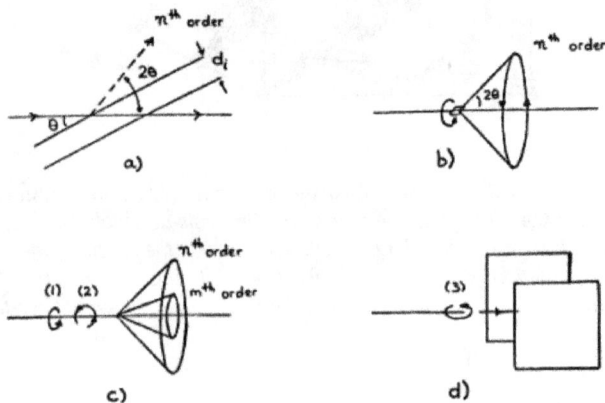

Figure 3. A schematic description of Debye Scherrer experiment.

These days, in practice, we assay the crystallinity of a powder sample and determine the size of the nano-crystallites by using sophisticated instrumentation and using monochromatic X-rays. The facility consists of an X-ray tube, a sample holder and an X-ray detector. X-rays are generated in a cathode ray tube by heating a filament to produce electrons which are accelerated towards a copper target when characteristic X-rays are produced. These are further filtered to yield mono chromatic X-rays. X-ray diffraction line broadening method of a particular Bragg peak for angle 2θ is used to analyse the crystallinity of a sample. For example see Fig 4.

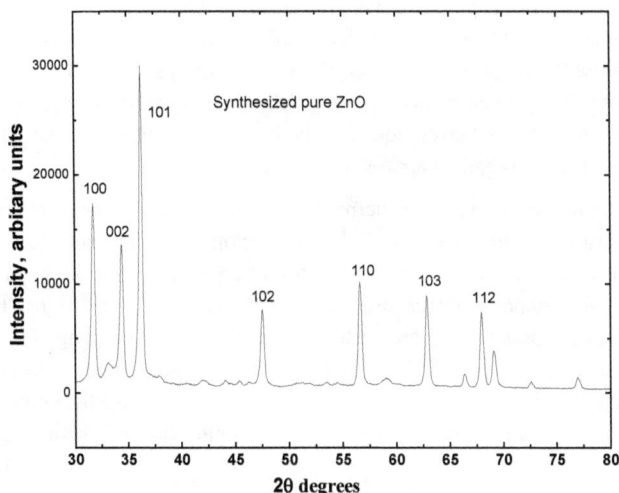

Figure 4. This shows the characteristic XRD patterns of the synthesized ZnO powder sample .

The phase structures of the samples has been identified by X-ray diffraction technique using Seifert XDAL 3000 diffractometer with CuKα radiation (wavelength of the radiation, k = 1.54 Å). The data have been collected in the range of (2θ) 30°–80° with a step size of 0.06°. Si has been used as external standard to deconvolute the contribution of instrumental broadening [1]. The XRD pattern has been shown in Fig. 4. The grain sizes of the synthesized samples have been calculated using the Scherrer formula [1]:

$$D_{hkl} = K\lambda/\beta_P \cos\theta \qquad\qquad (i)$$

where, D_{hkl} is the average grain size, K the shape factor (taken as 0.9), λ is the X-ray wavelength, β_P is the full width at half maximum (FWHM) intensity (here 101 peak of the ZnO spectrum fitted with a Gaussian, for precision measurement) after subtracting the instrumental line broadening, in radians. The appearance of characteristics diffraction peaks for pure ZnO corresponding to (1 0 0), (0 0 2), (1 0 1), (1 0 2), (1 1 0), (1 0 3) and (1 1 2) planes is in good agreement with the standard XRD peaks of crystalline bulk ZnO with hexagonal wurtzite structure [JCPDS card No. 36–1451, a = 3.2501 Å, c = 5.2071 Å, space group: P_63mc (1 8 6)]. Further, the average grain sizes of the ZnO samples were estimated from X-ray line broadening using Scherrer's equation. The particle size of pure ZnO is 41 nm. The Scherrer equation is limited to nano-scale particles. It is not applicable to grains larger than approx. 0.1 to 0.2 μm.

It is important to realize that the Scherrer formula provides a lower bound on the particle size. The reason for this is that a variety of factors can contribute to the width of a diffraction peak besides instrumental effects and crystallite size; the most important of these are usually inhomogeneous strain and crystal lattice imperfections. The following sources of peak broadening are listed as dislocations, stacking faults, twinning, microstresses, grain boundaries, sub-boundaries, coherency strain, chemical heterogeneities, and crystallite smallness. (Some of those and other imperfections may also result in peak shift, peak asymmetry, anisotropic peak broadening, or affect peak shape.)

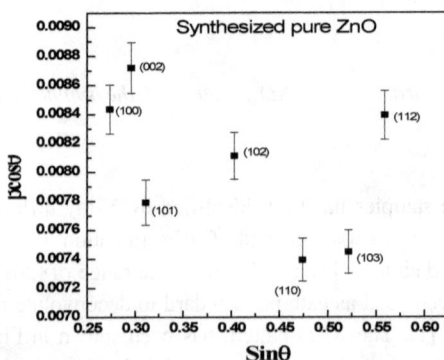

Figure 5. Williamson–Hall plots for pure ZnO nano crystalline sample, for structural strain analysis. Error bar indicates the structural broadening of particular peak.

If all of these other contributions to the peak width were zero, then the peak width would be determined solely by the crystallite size, then the Scherrer formula can be applied. If the other contributions to the peak broadening be present, then the size can be larger than the actual situation, with the "extra" peak width coming from the other factors. The concept of crystallinity can be used to collectively describe the effect of crystal size and imperfections on peak broadening.

As seen in the following, nano crystalline material usually suffers from structural strain as the grain interior is relatively defect free but the grain boundary consists of high-density defect clusters [2,3]. Thus, the strain in the lattice has been estimated through constructing Williamson–Hall (W–H) plot, with different Bragg peaks [4] taken into consideration, such as:

$$\beta \cos\theta = K\lambda / D_{hkl} + 2\varepsilon \sin\theta \qquad (ii)$$

where, ε is the micro strain parameter. One can estimate the strain in ZnO structure due to its size effect by W-H plot, shown in Fig. 5. Considerable anisotropy in structure has been noticed since unambiguous linear plot of the strain from all Bragg angles were not possible as seen from Fig. 5, the reason possibly lies with the surface effect of the crystallites. The results have been shown for only representative sample [for pure ZnO (~ 41 nm)] .

3.   Electron microscopy

An electron microscope is a microscope that uses a beam of accelerated electrons as a source of illumination. Electrons are charged particles (point charges with rest mass). The electron also has an associated spin of +1/2. While in motion an electron possesses kinetic energy, regardless of any imposed charge field—this could be achieved by accelerating electrons via a voltage differential into a screened "field-free" region, which initially imparts the energy required to accelerate the electron. Given sufficient voltage, the electron can be accelerated sufficiently fast to exhibit measurable relativistic effects, and the velocity must be accounted accordingly. According to the wave particle duality, electrons can also be considered as wave propagations and therefore have associated wave properties such as wavelength, phase and amplitude. As the wavelength of an electron can be up to 100,000 times shorter than that of visible light photons, electron microscopes have a higher resolving power than light microscopes and can reveal the structure of smaller objects. A scanning transmission electron microscope has achieved better than 50 pm resolution in annular dark-field imaging mode [6] and magnifications of up to about 10,000,000x whereas most light microscopes are limited by diffraction to about 200 nm resolution and useful magnifications below 2000x.

Electron microscopes have electron optical lens systems (electric and magnetic fields) that are analogous to the glass lenses of an optical light microscope.

Electron microscopes are used to investigate the ultra small structure of a wide range of biological and inorganic specimens including microorganisms, cells, large molecules, biopsy samples, metals, and crystals, detailed morphology of nano-materials. Modern electron microscopes produce electron micrographs using specialized digital cameras and frame grabbers to capture the image.

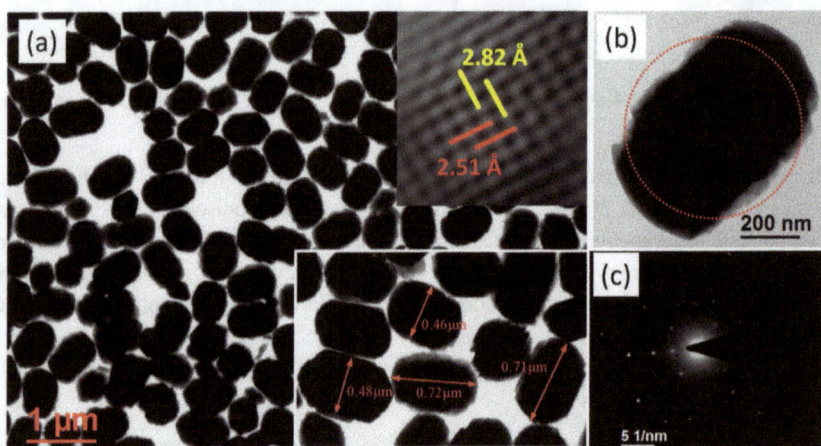

Figure 6. An illustrative example of TEM picture, a)showing Ga-oxy hydroxide nanoparticles, b) depicting their size and c)electron diffraction pattern (reference 7).

2.1 Transmission electron microscope (TEM)

Here in a simple manner is described the principle and some essentials of the technique. The original form of electron microscope, the transmission electron microscope (TEM) uses a high voltage electron beam to illuminate the specimen and create an image. The electron beam is produced by an electron gun, commonly fitted with a tungsten filament cathode as the electron source. The electron beam is accelerated by an anode typically at +100 keV (40 to 400 keV) with respect to the cathode, focused by electrostatic and electromagnetic lenses, and transmitted through the specimen that is in part transparent to electrons and in part scatters them out of the beam. When it emerges from the specimen, the electron beam carries information form the structure of the specimen that is magnified by the objective lens system of the microscope. The spatial variation in this information (the "image") may be viewed by projecting the magnified electron image onto a fluorescent viewing screen coated with a phosphor or scintillator material such as zinc sulfide. Alternatively, the image can be photographically recorded by exposing a photographic film or plate directly to the electron beam. A high-resolution phosphor may be coupled by means of a lens optical system or a fibre optic light-guide to the sensor of a digital camera. The image detected by the digital camera may be displayed on a monitor or computer.

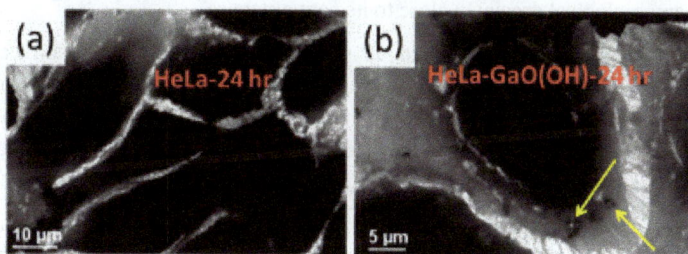

Figure 7. An illustration of biological specimen : Hela cells , a) alive in a nutritive solution , b) Nano particles loaded in the cell.

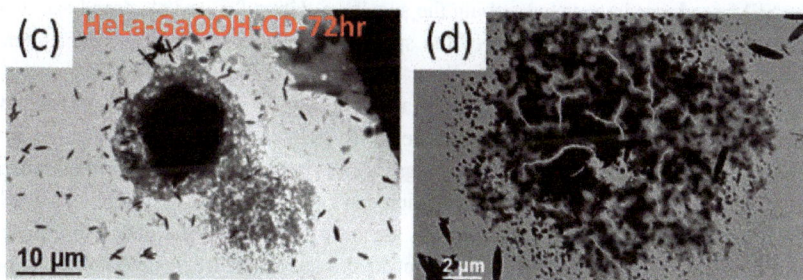

Figure 7. This picture shows the cell death after administration of cyclodextrin coated Gd –oxy hydroxyl nano particles (reference 7).

Sample preparation for nanomaterials

Nanomaterials are generally dispersed either in pure water/ isopropyl alcohol (in which they are insoluble) and mildly sonicated in a bath sonicator. Then the solution is put drop by drop (one or two drops only) in a tiny copper grid (mesh) coated with carbon and dried thoroughly in a vacuum desiccator. They should be moisture free or free from alcohol vapour. The specimen are placed in high vacuum chamber.

For biological samples special freeze drying process is adapted, Cryofixation – freezing a specimen so rapidly, in liquid ethane, and maintained at liquid nitrogen or even liquid helium temperatures, so that the water forms vitreous (non-crystalline) ice. This preserves the specimen in a snapshot of its solution state. An entire field called cryo-electron microscopy has branched from this technique. With the development of cryo-electron microscopy of vitreous sections (CEMOVIS), it is now possible to observe samples from virtually any biological specimen close to its native state.

Apart from these, conductive coating is sometimes needed: an ultrathin coating of electrically conducting material, deposited either by high vacuum evaporation or by low vacuum sputter coating of the sample. This is done to prevent the accumulation of static electric fields at the specimen due to the electron irradiation required during imaging. The coating materials include gold, gold/palladium, platinum, tungsten, graphite, etc.

Earthing – to avoid electrical charge accumulation on a conductive coated sample, it is usually electrically connected to the metal sample holder. Often an electrically conductive adhesive is used for this purpose.

2.2   Scanning electron microscope (SEM)

The SEM produces images by probing the specimen with a focused electron beam that is scanned across a rectangular area of the specimen . When the electron beam interacts with the specimen, it loses energy by a variety of mechanisms. The lost energy is converted into alternative forms such as heat, emission of low-energy secondary electrons and high-energy backscattered electrons, light emission (cathodoluminescence) or X-ray emission, all of which provide signals carrying information about the properties of the specimen surface, such as its topography and composition. The image displayed by an SEM maps the varying intensity of any of these signals into the image in a position corresponding to the position of the beam on the specimen when the signal has been generated.

In SEM images of a sample is produced by scanning the surface with a focused beam of electrons. It is not primarily used for determination of size of the nano-particles only. The electrons interact with atoms in the sample, producing various signals that contain information about the sample's surface topography and composition. Samples for SEM have to be prepared and dried to withstand the vacuum conditions and high energy beam of electrons, under this drying conditions, nanoparticles can highly aggregate

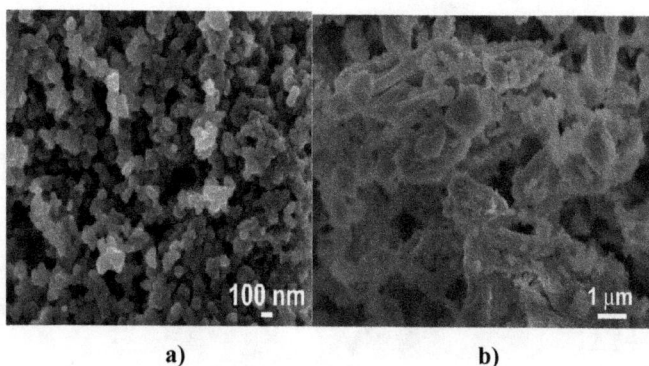

Figure 8. FE-SEM images of a)hydrated Ruthenium oxide nano particle sand b) L-cysteine conjugated Ruthenium oxide material (ref. 8).

In this case (as shown in Fig 8), low-voltage SEM is typically conducted in an FE-SEM because field emission guns (FEG) are capable of producing high primary electron brightness and small spot size even at low accelerating potentials. To prevent charging of non-conductive specimens, operating conditions has been adjusted such that the incoming beam current is equal to sum of out coming secondary and backscattered electrons currents, a condition that is more often met at accelerating voltages of 0.3–4 kV.

The SEM has compensating advantages, such as: the ability to image a comparatively large area of the specimen; the ability to image bulk materials (not just thin films or foils); and the variety of analytical modes available for measuring the composition and properties of the specimen.

Thus the purpose of using FE-SEM here has been to study the surface morphology modification of the conjugated –Ru oxide nano –material as compared to original hydrated Ru-oxide nano-material (Fig 8). Also it yields information about elemental composition.

2.3 Dynamic light scattering (DLS)

Dynamic light scattering (DLS), sometimes referred to as Quasi-Elastic Light Scattering (QELS), is a non-invasive, well-established technique for measuring the size and size distribution of molecules and particles typically in the submicron region, and with the latest technology lower than 1nm.

Typical applications of dynamic light scattering are the characterization of particles, emulsions or molecules, which have been dispersed or dissolved in a liquid. The Brownian motion of particles or molecules in suspension causes laser light to be scattered at different intensities. Analysis of these intensity fluctuations yields the velocity of the Brownian motion and hence the particle size using the Stokes-Einstein relationship.

Small particles in suspension undergo random thermal motion known as Brownian motion. This random motion is modelled by the Stokes-Einstein equation. Below the equation is given in the form most often used for particle size analysis

$$D_h = \frac{k_B T}{3\pi\eta D_t} \tag{iii}$$

The Stokes-Einstein relation that connects diffusion coefficient measured by dynamic light scattering to particle size.

Where D_h is the hydrodynamic diameter (this is the goal: particle size!), D_t is the translational diffusion coefficient (we find this by dynamic light scattering), k_B is Boltzmann's constant (we know this), T is thermodynamic temperature (we control this) η is dynamic viscosity (we know this).

The equation serves as important reminder about a few points. The first is that sample temperature is important, at it appears directly in the equation. Temperature is even more important due to the viscosity term since viscosity is a stiff function of temperature. Finally, and most importantly, it reminds the analyst that the particle size determined by dynamic light scattering is the hydrodynamic size. That is, the determined particle size is the size of a sphere that diffuses the way as nanoparticle.

(For work with protein sizing and other areas where hydrodynamic radius is more commonly used, note that the development here is around diameter. Radius calculations are the same except for a factor of two.)

How to Measure Particle Motion I: Dynamic Light Scattering Optical Setup

A top view of the optical setup for DLS is shown below.

The "noise" is actually due to particle motion an can be used to extract the particle size. In contrast to laser diffraction, DLS measurements are typically made at a single angle, although data obtained at several angles can be useful. In addition, the technique is completely non-invasive. The variations in the signal arise due to the random Brownian motion of the particles.

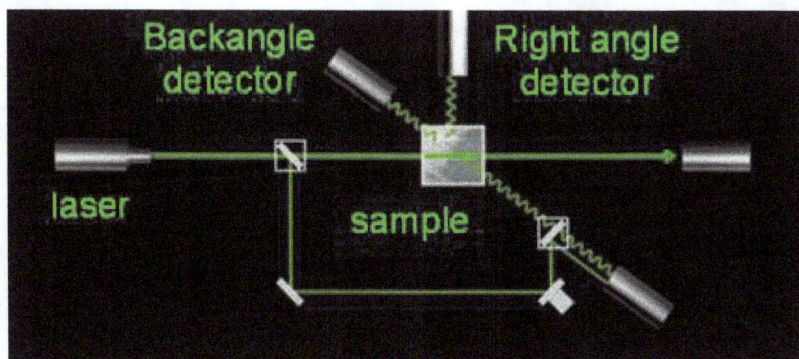

Figure 9. Instrimentation setup for DLS measurement. Light from the laser light source illuminates the sample in the cell. The scattered light signal is collected with one of two detectors, either at a 90 degree (right angle) or 173 degree (back angle) scattering angle. The provision of both detectors allows more flexibility in choosing measurement conditions. Particles can be dispersed in a variety of liquids. Only liquid refractive index and viscosity needs to be known for interpreting the measurement results.

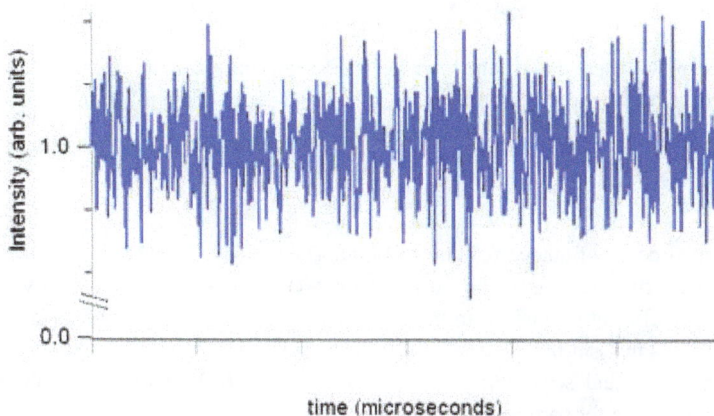

Figure 10. The obtained optical signal shows random changes due to the randomly changing relative position of the particles. This is shown schematically in the graph below.

How to Extract Particle Diffusion Coefficient: Dynamic Light Scattering Data Interpretation

The signal can be interpreted in terms of an autocorrelation function. Incoming data is processed in real time with a digital signal processing device known as a correlator and the autocorrelation function as a function of delay time, τ, is extracted.

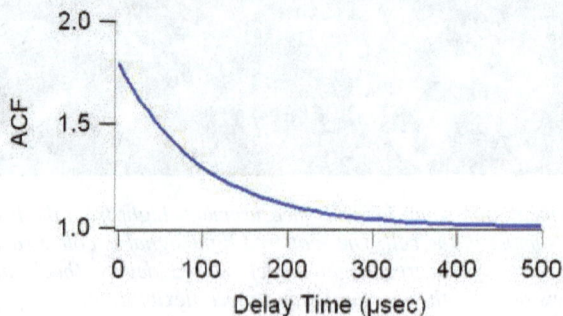

Figure 12. Autocorrelation Function from dynamic light scattering. The decay of this function is used to extract particle size. Faster decays correspond to smaller particles.

For a sample where all of the particles are the same size, the baseline subtracted autocorrelation function, C, is simply an exponential decay of the following form:

$$C = \exp(-2\Gamma\tau)$$ (iv)

Exponential decay of autocorrelation function. The decay constant is proportional to the diffusion coefficient. Γ is readily derived from experimental data by a curve fit. The diffusion coefficient is obtained from the relation $\Gamma = D_t q^2$ where q is the scattering vector, given by $q = (4\pi n/\lambda)\sin(\theta/2)$. The refractive index of the liquid is n. The wavelength of the laser light is λ, and scattering angle, θ. Inserting D_t into the Stokes-Einstein equation above and solving for particle size is the final step.

Analyzing Real Particle Size Distributions I: The Method of Cumulants and Z-average:

The discussion above can be extended to real nanoparticle samples that contain a distribution of particle sizes. The exponential decay is rewritten as a power series: the average diffusion coefficient and is used to extract average particle size; exponential decay of autocorrelation function. The linear decay constant is proportional to the particle size.

$$C = \exp(-2\bar{\Gamma}\tau + \mu_2 \tau^2 - \cdots)$$

(v)

Once again, a decay constant is extracted and interpreted to obtain particle size. However, in this case, the obtained particle size, known as the z-average size, is a weighted mean size. Unfortunately, the weighting is somewhat convoluted. Recall that the decay constant is proportional to the diffusion coefficient. So, by dynamic light scattering one has determined the intensity weighted diffusion coefficient. The diffusion coefficient is inversely proportional to size. So, in truth, the "z-average size" is the intensity weighted harmonic mean size. This definition differs substantially from that of the z-average radius of gyration encountered in the light scattering study of polymers.

Despite the convoluted meaning, the z-average size increases as the particle size increases. And, it is extremely easy to measure reliably. For these reasons, the z-average size has become the accepted norm for particle sizing by dynamic light scattering.

Analyzing Real Particle Size Distributions II: Size Distribution Data

It is possible to extract size distribution data from DLS data. One can convert the measured autocorrelation function into what is known as an electric field autocorrelation function, $g_1(\tau)$. Then use the following relationship between $g_1(\tau)$ and the scattered intensity, S, for each possible decay constant, Γ. The overall electric field autocorrelation function is the intensity weighted sum of the decays due to every particle in the system.

$$g_1(\tau) = \int S(\Gamma) \exp(-\Gamma \tau) d\Gamma$$

(vi)

Electric field autocorrelation function as a sum of exponential decays. The decay constants are inversely proportional to the particle size.

Inversion of this equation, that is using experimentally determined values of $g_1(\tau)$ to find values of $S(\Gamma)$, will lead to information about the size distribution.

All of these equations and the analysis are handled automatically in the software of the modernday instrumentation. As such, dynamic light scattering has found application for determining protein size, nanoparticle size, and colloid size. Consider for example ZnO sol, at physiological pH~7-7.5, hydrated with aging as shown in Fig 9.

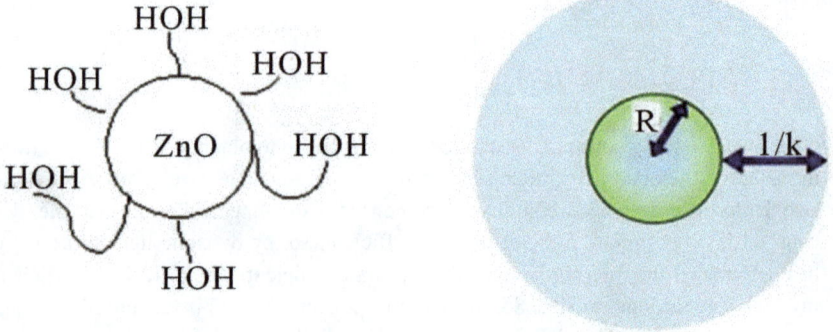

Figure 9. Schematic diagram of the hydrated layer of molecules (dipoles) arranged around the nanoparticle ZnO ('R' denote the radius) in TDW medium showing the Debye layer (1/κ).

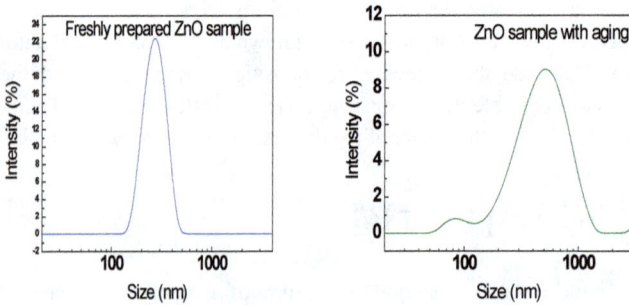

Figure 10. Example of determination of nano particle hydrodynamic size by dynamic light scattering method (Ref 9.)

Table 1: Comparison of the hydrodynamic size (2R) of synthesized pure ZnO and FA-ZnO samples Obtained from DLS

Sample	Size pure ZnO from TEM	Hydrodyna mic size of pure ZnO	Debye Layer	Size of FA-ZnO from TEM	Hydrodyna mic size FA-ZnO	Debye Layer FA-ZnO
Freshly prepared	~4 nm	388 nm	192 nm	~11 nm	518 nm	254 nm
After aging, ~ 1 month	-do-	560 nm	278nm	-do-	688 nm	388.5nm
After 2 months	-Do-	560nm	-Do-	-Do-	688nm	-do-

References

[1]   B.D. Cullity, S.R. Stock: *Elements of X-ray Diffraction*, Prentice-Hall, Englewood Cliffs, New Jersey, 2001.

[2]   T.E.M. Staab, R. Krause-Rehberg, B. Kieback: Review Positron annihilation in fine-grained materials and fine powders—an application to the sintering of metal powders, *J Mater Sci* 34 (1999) 3833-3851. https://doi.org/10.1023/A:1004666003732

[3]   T. Ungár, G. Tichy, J. Gubicza, *et al.*, Correlation between subgrains and coherently scattering domains., Powder Diffr 20 (2005) 366-375. https://doi.org/10.1154/1.2135313

[4]   GK Williamson, WH Hall: X-ray line broadening from field aluminium and Wolfram., Acta Metall 1(1953) 22-31. https://doi.org/10.1016/0001-6160(53)90006-6

[5]   Sreetama Dutta and Bichitra N. Ganguly, Characterization of ZnO nanoparticles grown in presence of Folic acid template, J. Nanobiotechnology 10 (2012) 29-38. https://doi.org/10.1186/1477-3155-10-29

[6]   R. Erni, M.D. Rossell, C. Kisielowski, U. Dahmen, Atomic-Resolution Imaging with a Sub-50-pm Electron Probe. Physical Review Letters. 102 (2009) 096101. https://doi.org/10.1103/PhysRevLett.102.096101

[7]   Bichitra Nandi Ganguly, Vivek Verma, Debanuj Chatterjee, Biswarup Satpati, Sushanta Debnath and Partha Saha, Study of Gallium Oxide Nanoparticles Conjugated with β-cyclodextrin -An Application to Combat Cancer, ACS Materials and Interfaces 8 (2016) 17127- 17137. https://doi.org/10.1021/acsami.6b04807

[8]   Bichitra Nandi Ganguly , Buddhadeb Maity, Tapan Kumar Maity, Joydeb Manna, Modhusudan Roy, Manabendra Mukherjee , Sushanta Debnath, Partha Saha, Nagaraju Shilpa, and Rohit Kumar Rana, l-Cysteine-Conjugated Ruthenium Hydrous Oxide Nanomaterials with Anticancer Active Application,   Langmuir 4 (2018) 1447-1456. https://doi.org/10.1021/acs.langmuir.7b01408

[9]   Sreetama Dutta and Bichitra Nandi Ganguly, Characteristics of Dispersed ZnO-Folic acid Conjugate in Aqueous Medium, Advances in Nanoparticles 3(2014) 23-30. https://doi.org/10.4236/anp.2014.31004

Chapter 6

Probing Defects by Positron Annihilation Spectroscopy

Mahuya Chakrabarti[1] and Dirtha Sanyal[2*]

[1]Department of Physics, Basirhat College, Basirhat, North 24 Parganas, Pin-743412, India

[2]Variable Energy Cyclotron Centre,1/AF, Bidhannagar, Kolkata-700064, India

*dirtha@vecc.gov.in

Abstract

Positron annihilation technique is a well-known nuclear solid state probe to characterize structural defects in a material/nano material. Since the nano size materials often suffer from vacancy type or surface structural defects, they are interesting materials for the nanotechnology/nano-bio platform. The defects and the chemical nature of the defects can be probed by the positron annihilation techniques which are applicable to biochemical scaffold as they often deal with surface chemical reactions in the biological perspectives. The positron annihilation lifetime (PAL) spectroscopy, Doppler broadening (DB) spectroscopy and the coincidence Doppler broadening (CDB) spectroscopy are the three useful positron annihilation techniques employed in different materials to characterize structural defects and that to understand the material itself in terms of its molecular organization. In the present article we will discuss the basics of the above three positron annihilation techniques and then by employing these techniques an important conclusion will be discussed in the field of wide band gap semiconductor oxide material, namely the "defect induced ferromagnetism". The growth and engineering of metallic oxide nano particles are useful in medicinal applications. Ferromagnetic property is important in the application of nanoparticles in diverse fields like biomedicine, where intensive research is currently being conducted, at least on one diagnostic application of magnetic nanoparticles as magnetic resonance imaging contrast agents.

Keywords

Positron Annihilation Technique, Defects, Semiconductors, Bio-applications

Contents

1. Introduction..124

2. Basics of the positron annihilation technique 124

3. The doppler broadening data analysis 132

    3.1 Line shape analysis 132

    3.2 Ratio-curve analysis..133

4. Results and discussions 133

5. Identifying the defects in annealed ZnO nano material 133

6. Defect-magnetism correlation in wide band gap semiconductor......137

Conclusion..138

References 138

1. Introduction

Nuclear solid state techniques, mainly positron annihilation techniques [1-3], Mössbauer spectroscopy [4] and perturbed angular correlation spectroscopy [5] are widely used to characterize different electronic structural properties of various materials. In the present article we are discussing the application of positron annihilation spectroscopy to characterize defects and thereby identifying the underlining physics in different oxide nano-size materials. Such techniques are often useful for determining subtle micro-structural changes in a given molecular substrate and are also applicable as multi-dimensional probe. The basic understanding of the technique may also lead to its applicability in diagnostic applications of bio-medicine [6] imaging technology.

2. Basics of the positron annihilation technique

In the last fifty years or more, positron annihilation technique has been advantageously employed to characterize defects in different solids [1,2,7,8]. Presently there are a number of positron facilities and a large number of experimental works have been reported by employing these techniques. In the conventional positron annihilation techniques a radioactive positron (β^+) emitting source like ^{22}Na, is used. The entire beta spectrum from the radioactive nuclei (Fig. 1) consists of different energetic beta particles, with maximum β^+ energy (E_{max}) of 540 keV in case of ^{22}Na. Entering into the solid, the energetic positrons become thermalized within 1 to 10 ps by producing electron-hole pairs and phonons and then diffuse (~ 100 nm) inside the material [2]. Depending upon the energy, positrons will have a different penetration depth, t, which is ~ $E^{1.4}/16\rho$, where, ρ is the density of the studied material. Thermalized positrons get highly localized when they are captured at defect sites, e.g., vacancies, small and large angle grain boundaries, dislocations, voids. etc. The eventual annihilation of the thermalized positron with an electron in the studied material is in general (379 out of 380) a two 511 keV - γ annihilation process in the centre of the mass frame.

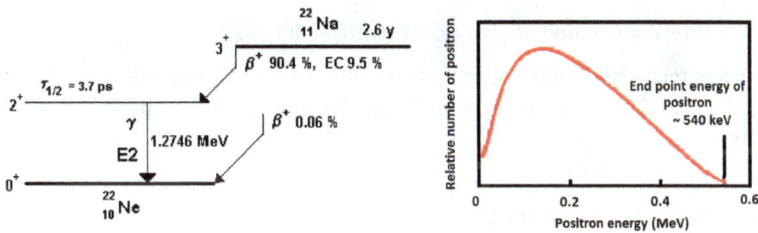

Figure 1. The decay scheme of ^{22}Na and the positron energy spectrum.

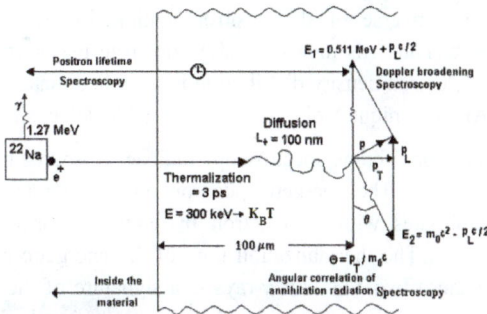

Figure 2. Schematic representations of the three basic positron annihilation techniques: (i) Positron annihilation lifetime (PAL) spectroscopy, (ii) Doppler broadening spectroscopy and (iii) Angular correlation of the annihilation radiation spectroscopy.

One can selectively choose the positrons, with a definite energy between E and E + ΔE, from the entire e$^+$ spectrum (E can be varied) and injected inside the studied material. In that case one can probe a precise thickness. This is known as depth resolved positron beam facility. There are mainly two type of positron beam facility (i) source based (^{22}Na) positron beam facility [9] and (ii) accelerator (electron-linac) [10] or reactor based positron beam facility [11]. The intensity in the source based positron beam facility is limited while one can have a high intensity ($\sim 10^6$ to 10^8 positrons per second) in the accelerator based beam facility.

Fig. 2 represents the schematics of three major positron annihilation techniques which are discussed below.

A. *Positron annihilation lifetime (PAL) spectroscopy*

In the non-relativistic limit positron annihilation rate, λ, is proportional to the overlap integral of the electron density $n_-(\mathbf{r})$ at the annihilation site and the positron density $n_+(\mathbf{r})$ = $|\Psi^+(\mathbf{r})|^2$ [1],

$$\lambda = 1/\tau = \pi r_0^2 c \int |\Psi^+(\mathbf{r})|^2 \gamma \, n_-(\mathbf{r}) \, d\mathbf{r} \qquad (1)$$

where r_0 (= e^2/m_0c^2) is the classical electron radius, c the speed of light, and $\mathbf{r}$ the position vector. γ, the enhancement factor ($\sim 1+\Delta n_-/n_-$), is the correlation function and it describes the increase of the electron density (Δn_-) due to the Coulomb attraction between electron and positron. Inverse of the positron annihilation rate, λ, is the positron annihilation lifetime, τ. Thus τ is inversely proportional to the electron number density and one can map the electron density distribution in a studied material by measuring the positron lifetime. PAL is a unique tool to study the possible defect sites in a material.

In the positron annihilation lifetime spectroscopy radioactive source like ^{22}NaCl has been used as a positron source. The energetic positron on entering inside a material, get thermalized and annihilates with an electron by emitting normally two oppositely directed 511 keV γ-rays. The thermalization time of the energetic positron is typically within 3 ps. One of these two 511 keV γ-rays is a signature of the annihilation of the positron with an electron and hence considered as the death signal of the positron. After 3.7 ps of the emission of β^+ from ^{22}Na source, the daughter nucleus, ^{22}Ne, de-excites to the ground state by emitting a γ-ray of energy 1.274 MeV, which is considered as the birth signal of the positron. The timing interval between the birth signal (1.274 MeV γ-ray) and the death signal (511 keV γ-ray) is considered as the lifetime of the positron inside the material.

For positron annihilation lifetime spectroscopic (PAL) study, about 10 µCi (micro-Ci) ^{22}NaCl source sealed in 1.5 µm thick nickel foil has been used as a positron source [12].

The sealed source has been sandwiched between two identical plane faced samples. Conventional fast-fast coincidence assembly using two gamma ray detectors (25 mm long and 25 mm tapered to 13 mm diameter BaF_2 scintillator optically coupled with XP2020Q photomultiplier tube) have been used for positron lifetime measurement system (Fig. 3). The spectrometer has a timing resolution (full width at half maximum) of ~ 220 ps measured by the prompt gamma ray of ^{60}Co source with the proper energy window (700–1320 keV for the start channel and 300–550 keV for the stop channel) [13].

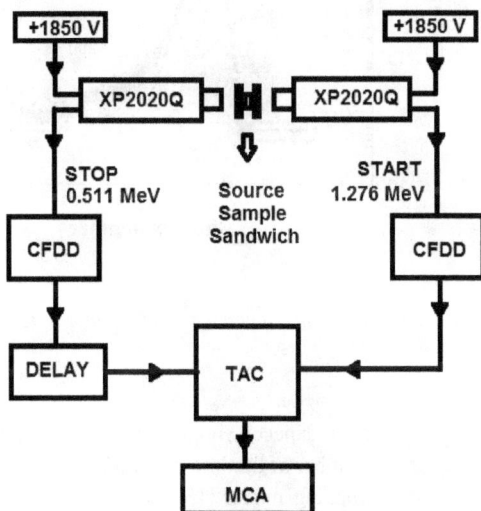

Figure 3. A typical γ–γ coincidence (fast-fast) setup for positron lifetime measurement

The positron lifetime component τ_i and its relative intensity I_i can be evaluated from the measured positron annihilation lifetime spectrum N (t) vs. t (Fig. 4) either by using the least square method or by the integral transform method. In the integral transform method, the lifetime spectrum is considered as the Laplace transform and the lifetime component is obtained by its inverse. The positron lifetime components have been evaluated here by using the widely used least square method. To evaluate the lifetime parameters from the measured spectrum computer programs (PATFIT – 88) [14] have been used widely.

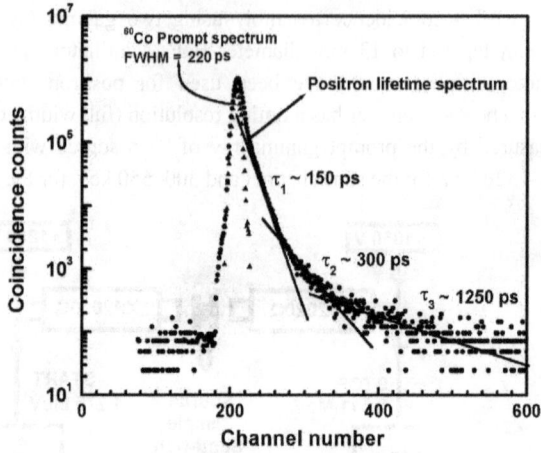

Figure 4. A typical positron lifetime spectrum. The prompt time resolution of the system using ^{60}Co is also shown in the figure.

In case of single crystal one can expect a single positron lifetime component whereas for polycrystalline samples the number of lifetime states will be different depending upon the material. In the two state trapping model [15] it is assumed that the smallest lifetime component, τ_1, is due to the positron annihilation in the bulk of the material and the longer lifetime component, τ_2, is due to the positron annihilation at the defect sites in the material. The annihilation lifetime in the Bloch state (called τ_B) is calculated as

$$\tau_B = (I_1 + I_2)/ (I_1/ \tau_1 + I_2 / \tau_2) \tag{2}$$

Therefore, in the two state trapping model τ_2 and τ_B are the physically significant quantities. The mean positron lifetime (τ_{av}) in terms of τ_i and I_i is calculated by the equation

$$\tau_{av} = \sum \tau_i I_i \tag{3}$$

B.   Angular correlation of annihilation radiation (ACAR) spectroscopy

Again due to non-zero momentum of the electron-positron pair, the annihilated γ photons deviate from collinearity in the laboratory frame by an angle [16]

$$\theta_{x, y} = p_{x, y} / (m_0 c) \tag{4}$$

where $p_{x, y}$ is the momentum component perpendicular to the propagation direction, m_0 is the rest mass of the electron. These equations hold good for even small angles. By a coincidence measurement $\theta_{x, y}$ can be measured simultaneously along both x and y direction. This is the basic of the angular correlation of annihilation radiation measurement technique. Thus, by measuring the angular correlation between the two annihilating gamma photons, one can also study the momentum distribution of the electrons in a solid.

C.   Doppler broadening of the electron positron annihilation γ-radiation (DBPAR) line-shape spectroscopy

99.7 % positrons are annihilation via two photon annihilation process. The annihilation results two oppositely directed (exactly 180° apart, in the center of mass frame) 511 keV γ-photon emission. Since the rest mass of positron or the electron is $m_0 c^2 = 511$ keV. But, before annihilation the electron-positron pair has some momentum, p, which is entirely due to the momentum of the electron, the positron is thermalized and hence its momentum is almost negligible (~ meV). During the annihilation process, the momentum of the electron-positron pair (p) is transferred to the photon pair to conserve the momentum. As a result of which the 511 keV annihilation γ- rays are Doppler shifted [2] by an amount $\pm \Delta E$ in the laboratory frame. Where

$$\pm \Delta E = p_L c/2 \tag{5}$$

p_L (pcosθ) is the component of the electron momentum, p, along the direction of the detection of the annihilating γ- rays. Fig. 5 represents the Doppler shift of the electron positron pair along the detector direction due to non-zero momentum of the electron positron pair. Thus the 511 keV energy line is broadened due to the individual Doppler shifts of large numbers of annihilation events and is called the Doppler broadening spectrum. Hence the Doppler broadened spectrum maps the momentum distributions of the electrons at the positron annihilation site.

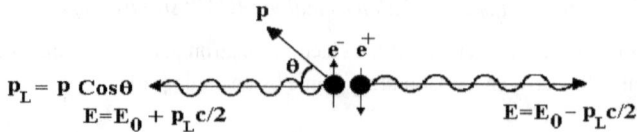

$$p_L = p \cos\theta$$
$$E = E_0 + p_L c/2$$
$$E = E_0 - p_L c/2$$

Figure 5. Schematic representation of Doppler shift of the annihilating γ- rays along the detector direction.

D. Coincidence Doppler broadening spectroscopy

The wing part of the Doppler broadened 511 keV γ-ray spectrum contains the information of positron annihilation with the higher momentum electron such as core electron of the atom. In a real experimental setup these wing parts (having very low counts compared to the photo peak counts) are largely affected due to the Compton part of the 511 keV γ-ray as well as the associated 1276 keV γ-ray (which is coming from the decay of $^{22}Ne_{10}$ isotope) in case of $^{22}NaCl$ positron source, which act as background counts. Normally, using a single HPGe detector the ratio between the photo-peak counts to the wing part counts are 300:1. Normally the electron positron annihilation results in two oppositely directed 511 keV γ-rays. Now instead of a single HPGe detector if one uses two identical HPGe detectors (placed oppositely i.e., 180° apart) in coincidence the Compton part of the 1276 keV γ-rays can be suppressed and the peak to background ratio can be improved more than an order (10000:1). In the above configuration using the ±ΔE selection the peak to background can be improved to 10^5:1 [17]. ± ΔE selection means for a particular electron positron annihilation if the pair has a momentum component along the detector axis is p_L (pcosθ) then one 511 keV γ-ray will be Doppler shifted to 511 + ΔE keV and the other will be 511 - ΔE keV, where ΔE = $p_L c/2$ [2]. Fig. 6 represents the above mentioned coincidence Doppler broadened (CDB) spectrometer and Fig. 7 is a typical CDB spectrum. A comparison of 511 keV γ-ray spectrum recorded in a single HPGe detector; two identical HPGe detector in coincidence; and extracted coincidence Doppler broadening spectrum with ± ΔE selection has been plotted in Fig. 8. The application of this almost background free Doppler broadening spectrum is most sensitive positron annihilation spectroscopy to identify the chemical nature of the defect. Using CDB spectroscopy a large number of defects related physical and chemical properties of materials have been reported in last several years [18-29].

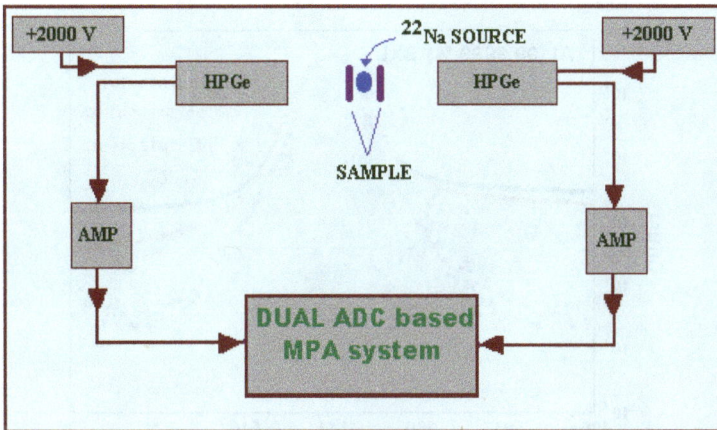

Figure 6. Schematic representation of coincidence Doppler broadening spectrometer using two identical HPGe detectors.

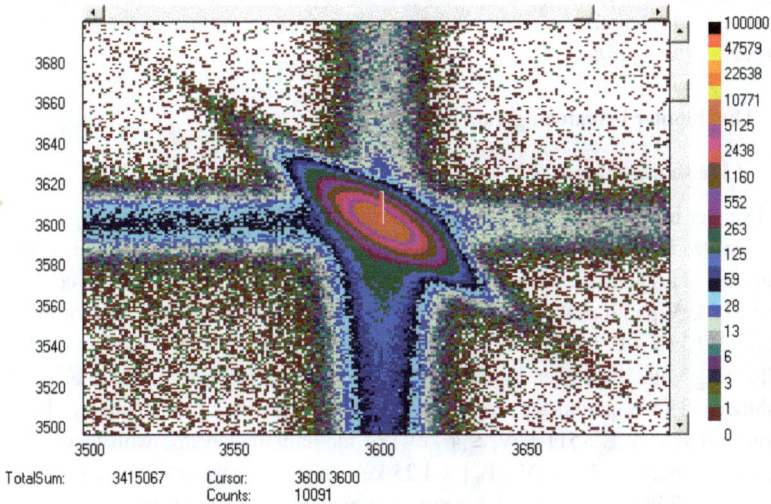

Figure 7. A typical coincidence Doppler broadening spectrum using dual ADC based multiparameter data acquisition system.

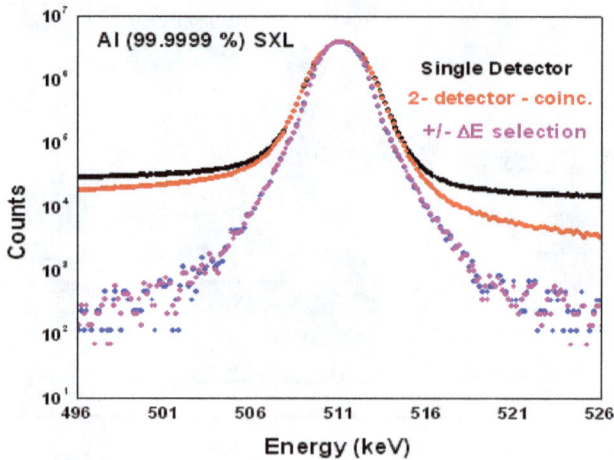

Figure 8. The 511 keV gamma ray line recorded in three possible set-ups (i) using a single HPGe detector; (ii) using two HPGe detectors with standard coincidence technique; (iii) using the concept of ± ΔE selection along with the two HPGe detectors coincidence configuration.

3.   The doppler broadening data analysis

3.1   Line shape analysis

The Doppler broadening of the electron positron annihilated 511 keV γ-ray spectrum has been analyzed by evaluating the so called line-shape parameters (S-parameter and W-parameter) [1]. The S-parameter is calculated as the ratio of the counts in the central area of the 511 keV photo peak (| 511 keV - E_γ | ≤ 0.85 keV) and the total area of the photo peak (| 511 keV - E_γ | ≤ 4.25 keV). The S-parameter represents the fraction of positron annihilating with the lower momentum electrons with respect to the total electrons annihilated. The W-parameter represents the relative fraction of the counts in the wings region (1.6 keV ≤ |E_γ -511 keV| ≤ 4 keV) of the annihilation line with that under the whole photo peak (| 511 keV - E_γ | ≤ 4.25 keV). The W-parameter corresponds to the positrons annihilating with the higher momentum electrons. Fig. 9 represents the pictorial depiction of S and W parameter. S-parameter represents the relative defects in a system.

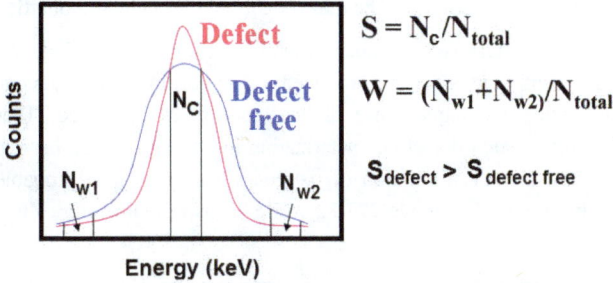

$$S = N_c/N_{total}$$

$$W = (N_{w1}+N_{w2})/N_{total}$$

$$S_{defect} > S_{defect\ free}$$

Figure 9. Pictorial depiction of line shape parameters (S & W).

3.2   Ratio-curve analysis

To identify the contributions of the valence and the core electron momentum involved in the annihilation process ratio curve analysis [30, 31] have been followed. Ratio-curve is defined as point to point ratio of area normalized CDB spectrum of the material under study with an area normalized CDB spectrum of reference sample. Normally a vacuum annealed 99.9999% pure Al single crystal or some other defect free samples have been used as a reference sample.

4.   Results and discussions

In this section, some applications of positron annihilation spectroscopy have been discussed to show the uniqueness of these techniques.

5.   Identifying the defects in annealed ZnO nano material

In oxide semiconductors like ZnO, MgO, TiO_2 and other related compounds, several defects configurations offer different interesting defect induced phenomenon. Among them ZnO received much more attention due to its potential applications as an electronics material, especially in context of possible use as room temperature magnetic semiconductor [32,33] optoelectronic material [34], future generation solar cell [35] and in biological applications as well [36,37]. Even at room temperature ZnO can be used as ultraviolet luminescent material. ZnO received a lot of research attention in last 10 years not just because of its potential application as a diluted magnetic semiconductor but also the experimental observation of defect induced room temperature ferromagnetism

[38,39,18,19], the so called d^0 ferromagnetism in it. It is very interesting that defects play the most crucial role in determining the magnetic as well as optical properties of ZnO.

The annealing temperature is very much crucial in case of ZnO to induce defects in it. Both Zn vacancy and O vacancy are produced in ZnO during heat treatment. Employing positron annihilation technique these defects can be characterized. There are large numbers of report of such defect characterization in ZnO. In our earlier reports [18,40] we have observed that the formation of oxygen vacancy is most probable when ZnO samples have been annealed above 773 K, while at lower temperature Zn vacancies are most dominating.

In this particular work two different annealing temperatures, 573 K and 1073 K, has been chosen to generate two different kinds of vacancies in ZnO. Employing temperature dependent (90 K to 300 K) positron annihilation Doppler broadening spectroscopy [1,2] these two type of same "as received" ZnO samples have been characterized.

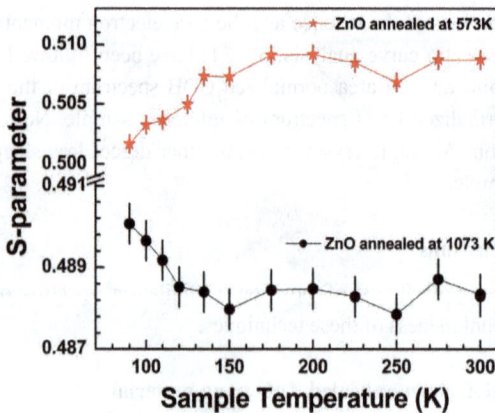

Figure 10. Temperature dependent S-parameter for two differently annealed ZnO.

Fig. 10 represents the temperature dependent Doppler broadening line shape parameter, *S*- parameter for both the samples. The absolute values of the S- parameter for the 1073 K annealed sample is little lower than the 573 K annealed samples. In earlier studies similar values of S- parameter at room temperature for the 573 K and 1073 K annealed ZnO samples has been observed. It is interesting that the S-parameter decreases with decreasing temperature for the 573 K annealed sample while for 1073 K annealed sample S- parameter increases with decreasing temperature. Decrease of S-parameter indicates that positrons are annihilating more with the higher momentum electrons with respect to lower momentum electrons. Thus the present study indicate that for 573 K annealed samples positrons are more annihilating with the higher momentum electrons while decreasing its temperature below 200 K. Tuomisto et al [41] have observed an increase of average positron lifetime with decreasing sample temperature below 200 K for their annealed ZnO sample, and conclude that this is due to the presence of vacancy defects which are negatively charge state (oxygen vacancy). They also observed a decrease of average positron lifetime due to decrease of sample temperature below 200 K for the as grown ZnO sample. In the positron annihilation spectroscopic technique nature of the variation of Doppler broadening S-parameter and the variation of average positron lifetime is almost same. The present work is in agreement with the earlier experimental observation [41,18]. In agreement with our earlier observations [18,42] Zn vacancy type defect are predominant in the "as received" ZnO sample while annealing above 500° C (773 K) oxygen vacancy type defects are generated in ZnO. The present 573 K annealed ZnO is Zn vacancy rich sample while the 1073 K annealed sample is oxygen vacancy reach sample. The nature of the temperature dependent S-parameter for two differently annealed ZnO samples (only annealing at two different elevated temperatures) is completely different below 200 K.

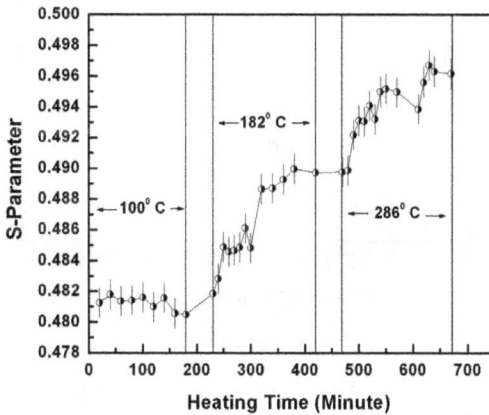

Figure 11. Variation of S-parameter with heating time during in-situ DBEPARL measurement of polycrystalline ZnO. (Figure taken from ref. Mat Research Exp.)

In-situ measurement of S-parameter during annealing upto 286° C of ZnO has been done for the 1st time and reported in ref. [43]. In this in-situ experiment a time sequential data has been recorded for each fixed temperature to understand the defect formation / saturation time. Fig. 11 represents the in-situ S-parameter vs. time for three steps of temperature. It is clear from the figure that annealing at 100°C upto 180 minutes the values of S-parameter remains almost constant indicating no change in defects in ZnO. Further increasing the annealing temperature to 182 °C S-parameter increases and gets a saturated value after 90 minutes. The results can be interpreted removal [44] of loosely bound hydroxyl (-OH) and carbon related species from ZnO which in a way creates some open volume defects as a result S-parameter increases. S-parameter increases further due to annealing at 286°C. Here also the saturation takes almost similar time. In this region defects migration and agglomeration process happens to form a defect cluster. The most probable defects are $V_{Zn}-V_O$, $2V_{Zn}-V_O$ and $V_{Zn}-2V_O$. Also the defect recovery process starts during annealing at 286°C since the recovery of isolated Zn vacancy is predicted at 266°C. [45]

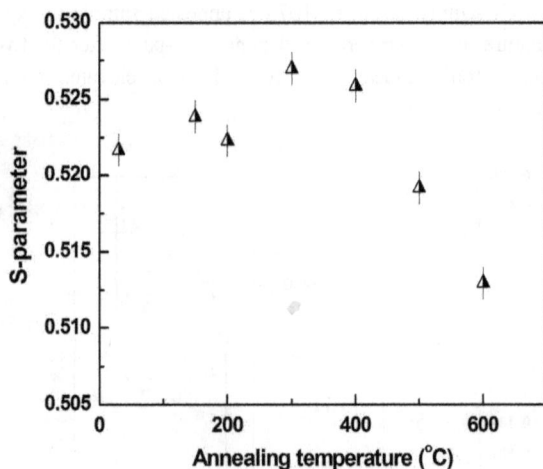

Figure 12. Variation of S-parameter with annealing temperature in ZnO.

In another experiment "as received" ZnO has been annealed at different temperature for 5 hour duration and after that the sample has been probed by off-line (at room temperature)

Doppler broadening spectroscopy to characterize the defect dynamics. Fig. 12 represents the S-parameter vs. annealing temperature graph. This shows that annealing beyond 400° C the S-parameter decreases drastically. This can be correlated with the oxygen loss from the ZnO beyond 500° C annealing.

6. Defect-magnetism correlation in wide band gap semiconductor

After the experimental observation by Sundaresan et al. [38] that the non-magnetic oxides shows room temperature ferromagnetism in its nano-crystalline state, positron annihilation techniques have extensively employed to establish the defect magnetism correlation. It is the positron annihilation techniques which established the defect induced ferromagnetism in the nano-crystalline non-magnetic oxides like ZnO, TiO$_2$, MgO etc. [18-20,23-25,46,47]

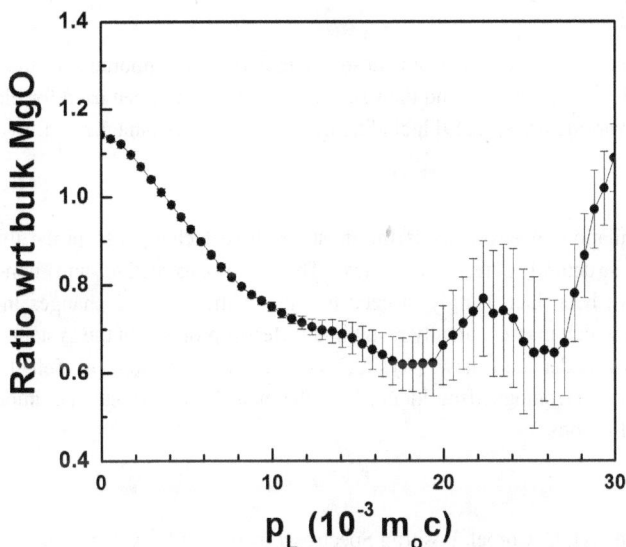

Figure 13. Area normalized ratio curve of nanocrystalline MgO with respect to bulk MgO.

Figure 13 represent the typical ratio curve. Here coincidence Doppler broadening spectra of nanocrystalline and bulk MgO have been recorded in the above said CDB setup. The ratio between the area normalized CDB spectra of nanocrystalline MgO with respect to bulk MgO have been plotted in the Fig. 13. It shows a broad dip at momentum value $p_L \sim 17 \times 10^{-3}$ m_oc. Using Virial theorem approximation (in the atom the expectation value of the kinetic energy of an electron, E_{kin}, is equal to the binding energy of the electron), one can calculate the E_{kin} using $p_L=(2\ m_0 E_{kin})^{1/2}$[30]. The energy of the 2p and 1s core electrons of Mg, is 50 eV and 88 eV respectively. The corresponding p_L will be $\sim 14 \times 10^{-3}$ m_oc and 18×10^{-3} m_oc. Thus the dip around the region $\sim p_L \sim 17 \times 10^{-3}$ m_oc as observed by Fig. 13 clearly shows the less annihilation of positrons with the core electrons of Mg, which suggests the presence of Mg vacancy in nanocrystalline MgO. Magnetic measurement shows the presence of room temperature ferromagnetism in these nanocrystalline MgO [20]. Thus from positron annihilation spectroscopy, mainly CDB spectroscopy the reason behind the room temperature ferromagnetism in nanocrystalline MgO has been identified as Mg vacancy. Similarly, in other oxides like ZnO [18], TiO_2 [46], SnO_2 [47] etc., employing positron annihilation techniques the vacancy defects have been identified.

It is pertinent mention here that such nano-oxide materials are important for the interface of biology and material science and their surface structure have been used for biomedical application in various ways[48,49] including the property of ferromagnetism.

Conclusion

Positron annihilation spectroscopy is the most sensitive technique to probe the atomic vacancy and structural defects in solids. These non-destructive and non-invasive techniques have been increasingly utilized to capture the essential changes in a given nano-material, which may eventually result in a different property in the system. The vast results help to understand new physics, one of such example is defect induced ferromagnetism. Ferromagnetism induced in the nano-materials can be important in biological applications.

References

[1]   P. Hautojarvi, C. Corbel, Positron Spectroscopy of Solids, A. Dupasquier, A. P. Mills Jr., (Eds.), IOS Press, Ohmsha, Amsterdam, 1995.

[2]   R. Krause-Rehberg and H. S. Leipner (Eds.), Positron Annihilation in Semiconductors, Springer Verlag, Berlin, 1999. https://doi.org/10.1007/978-3-662-03893-2

[3]   D.M. Schrader andY.C. Jean , Positron and Positronium Chemistry, Studies in Physical and Theoretical Chemistry, vol57, Elsevier, Amsterdam, 1988.

[4]   K. Routray, D. Sanyal and D. Behera, Dielectric, magnetic, ferroelectric, and
        Mossbauer properties of bismuth substituted nanosized cobalt ferrites through
        glycine nitrate synthesis method, J. of Appl. Phys. 122 (2017) 224104.
        https://doi.org/10.1063/1.5005169

[5]   Bichitra Nandi Ganguly, Sreetama Dutta , Soma Roy, Jens Röder , Karl Johnston,
        Manfred Martin, ISOLDE-Collaboration, Investigation on structural aspects of
        ZnO nano-crystal using radio-active ion beam and PAC , Nuclear Instruments and
        Methods in Physics Research B 362 (2015) 103–109.
        https://doi.org/10.1016/j.nimb.2015.08.098

[6]   Bichitra Nandi Ganguly, Nagendra Nath Mondal, Maitreyee Nandy, Frank
        Roesch, Some Physical Aspects of Positron Annihilation Tomography: a critical
        review; Journal of Radioanalytical and Nuclear Chemistry 279 (2009) 685-698.
        https://doi.org/10.1007/s10967-007-7256-2

[7]   P. Hautojarvi (Eds.), Positron in Solids, Springer-Verlag, Berlin, 1979.
        https://doi.org/10.1007/978-3-642-81316-0

[8]   W. Brandt and A. Dupasquier (Eds.), Positron Solid State Physics, North-Holland,
        Amsterdam, 1983.

[9]   R. S. Brusa, A. Dupasquier, R. Gfisenti, S. Liu, S. Oss and, A. Zecca, Deep
        disorder in neon-implanted copper single crystals detected by variable-energy
        positrons. J. Phys.: Condens. Matter 1 (1989), 5411-5420.
        https://doi.org/10.1088/0953-8984/1/32/010

[10]   T. Yamazaki, R. Suzuki, T. Ohdaira, T. Mikado and Y. Kobayashi, Production and
        application of pulsed slow positron beam using an electron linac, Radiation
        Physics and Chemistry 49 (1997) 651-659. https://doi.org/10.1016/S0969-
        806X(97)00015-7

[11]   C. Hugenschmidt, G. Kogel, K. Schreckenbach, P Sperr, M. Springer, B. Straßer,
        W.Triftshäuser, High intense positron beam at the new Munich research reactor
        FRM-II Appl. Sur. Sci. 149 (1999) 7-10. https://doi.org/10.1016/S0169-
        4332(99)00163-4

[12]   D. Sanyal, D. Banerjee, Udayan De, probing
        (Bi0.92Pb0.17)2Sr1.91Ca2.03Cu3.06O10+δ superconductors from 30 to 300 K
        by positron-lifetime measurements, Phys. Rev. B 58 (1998) 15226-15230.
        https://doi.org/10.1103/PhysRevB.58.15226

[13]   A. Sarkar, M. Chakrabarti, S. K. Roy, D. Bhowmick and D. Sanyal, Positron annihilation lifetime and photoluminescence studies on single crystalline ZnO. J of Phys. Cond. Matt. 23 (2011), 155801. https://doi.org/10.1088/0953-8984/23/15/155801

[14]   P. Kirkegaard, N. J. Pedersen and M. Eldrup, Report of Riso National Lab, (Riso-M-2740), 1989.

[15]   W. Brandt, in Positron Annihilation, edited by A. T. Stewart and L. O. Roellig (Academic, New York, 1967), p. 155. https://doi.org/10.1016/B978-0-12-395497-8.50014-X

[16]   J. Arponen and E. Pajanne, Angular correlation in positron annihilation. Journal of Physics F: Metal Physics 9 (1979), 2359-2376. https://doi.org/10.1088/0305-4608/9/12/009

[17]   K.G. Lynn, A.N. Goland, Observation of high momentum tails of positron-annihilation lineshapes, *Solid State Commun* 18 (1976) 1549-1552. https://doi.org/10.1016/0038-1098(76)90390-2

[18]   D. Sanyal, M. Chakrabati, T. K. Roy and A. Chakrabarti, The origin of ferromagnetism and defect-magnetization correlation in nanocrystalline ZnO. Phys. Letts. *A* 371(2007),482-485. https://doi.org/10.1016/j.physleta.2007.06.050

[19]   S Dutta, S Chattopadhyay, A Sarkar, M Chakrabarti, D Sanyal and D Jana, Grain size dependence of optical properties and positron annihilation parameters in Bi2O3 powder, Prog. Mater. Sci. 54 (2009), 89-136. https://doi.org/10.1016/j.pmatsci.2008.07.002

[20]   N. Kumar, D. Sanyal, and A. Sundaresan, Defect induced ferromagnetism in MgO nanoparticles studied by optical and positron annihilation spectroscopy Chem. Phys. Lett. 477 (2009), 360-364. https://doi.org/10.1016/j.cplett.2009.07.037

[21]   M. Chakrabarti, S. Dutta, S. Chattapadhyay, A. Sarkar, D. Sanyal and A. Chakrabarti, Grain size dependence of optical properties and positron annihilation parameters in Bi_2O_3 powder, Nanotechnology 15 (2004), 1792-1796. https://doi.org/10.1088/0957-4484/15/12/017

[22]   D. Sanyal, D. Banerjee, R. Bhattacharya, S. K. Patra, S. P. Chaudhuri, B. Nandi Ganguly and U. De, Study of transition metal ion doped mullite by positron annihilation techniques J. Mat. Sci. 31 (1996), 3447-3451. https://doi.org/10.1007/BF00360747

[23]   A. Sarkar, M. Chakrabarti, S. K. Roy, D. Bhowmick and D. Sanyal, Positron annihilation lifetime and photoluminescence studies on single crystalline ZnO, J of Phys. Cond. Matt. 23 (2011), 155801. https://doi.org/10.1088/0953-8984/23/15/155801

[24]   A. Sarkar, M. Chakrabarti, D. Sanyal, D. Bhowmick, S. Dechoudhury, A. Chakrabarti T. Rakshit and S. K. Ray, Photoluminescence and positron annihilation spectroscopic investigation on a H(+) irradiated ZnO single crystal. J. Phys.: Condens. Matter 24 (2012), 325503 (9 pages).

[25]   R.V.K. Mangalam, M. Chakrabrati, D. Sanyal, A. Chakrabarti, and A. Sundaresan, Identifying defects in multiferroic nanocrystalline BaTiO$_3$ by positron annihilation techniques. J of Phys. Cond. Matt. 21 (2009), 445902-445906. https://doi.org/10.1088/0953-8984/21/44/445902

[26]   U. De, D. Sanyal, S. Chaudhuri, P. M. G. Nambissan, Th. Wolf and H. Wuhl, Probing single-crystalline YBa2Cu3O7 across the superconducting transition temperature by positron annihilation measurements. Phys. Rev. B 62 (2000) 14519-14523. https://doi.org/10.1103/PhysRevB.62.14519

[27]   S. N. Guin, D. Sanyal and K. Biswas, The effect of order–disorder phase transitions and band gap evolution on the thermoelectric properties of AgCuS nanocrystals. Chem. Sci. 7 (2016), 534-543. https://doi.org/10.1039/C5SC02966J

[28]   S. N. Guin, S. Banerjee, D. Sanyal, S. K. Pati, and K. Biswas, Origin of the Order–Disorder Transition and the Associated Anomalous Change of Thermopower in AgBiS$_2$ Nanocrystals: A Combined Experimental and Theoretical Study Inorganic Chem. 55 (2016),6323-6331.

[29]   J. Dhar, S. Sil, A. Dey, P. P. Ray and D. Sanyal, Positron Annihilation Spectroscopic Investigation on the Origin of Temperature-Dependent Electrical Response in Methylammonium Lead Iodide Perovskite. J of Phys. Chem. Lett. 8 (2017), 1745-1751. https://doi.org/10.1021/acs.jpclett.7b00446

[30]   U Myler and P J Simpson, Survey of elemental specificity in positron annihilation peak shapes Phys. Rev. B 56 (1997), 14303-14309. https://doi.org/10.1103/PhysRevB.56.14303

[31]   P. Asoka-Kumar, M. Alatalo, V.J. Ghosh, A.C. Kruseman, B. Nielsen, K.G. Lynn, Increased Elemental Specificity of Positron Annihilation Spectra. Phys. Rev. Lett. 77 (1996), 2097-2100. https://doi.org/10.1103/PhysRevLett.77.2097

[32]   S. A. Wolf, D. D. Awschalom, R. A. Buhrman, J. M. Daughton, S. Von Molnar, M. L. Roukes, A. Y. Chtchelkanova and D. M. Treger, Spintronics: a spin-based electronics vision for the future.   Science 294 (2001) 1488-1495. https://doi.org/10.1126/science.1065389

[33]   P. Sharma, A. Gupta, K. V. Rao, F. J. Owens, R. Sharma, R. Ahuja, J. M. Osorio Guillen, B. Johansson and G. A. Gehring Ferromagnetism above room temperature in bulk and transparent thin films of Mn-doped ZnO Nat. Mater. 2 (2003), 673-677. https://doi.org/10.1038/nmat984

[34]   D. C. Kundaliya, S. B. Ogale, S. E. Loflanf, S. Dhar, C. J. Metting, S. R. Shinde, Z. Ma, B. Varughese, K. V. Ramanujachary, L. Salamanca –Riba and T. Venkatesan, On the origin of high-temperature ferromagnetism in the low-temperature-processed Mn-Zn-O system. Nat. Mater. 3 (2004), 709-714. https://doi.org/10.1038/nmat1221

[35]   J. Zhang, R. Skomski and D. J. Sellmyer, Sample preparation and annealing effects on the ferromagnetism in Mn-doped ZnO J. of Appl. Phys. 97 (2005), 10D303.

[36]   Sreetama Dutta , Sourav Sarkar and Bichitra Nandi Ganguly, Positron Annihilation Study of ZnO Nanoparticles Grown Under Folic Acid Template, J. Material Sci. Eng. 3(2014), 1000134.

[37]   Sreetama Dutta and Bichitra N Ganguly, Characterization of ZnO nano particles grown in presence of Folic Acid template, J. Nanobiotechnology 10:29 (2012)10 pages.

[38]   A. Sundaresan , R. Bhargavi, N. Rangarajan, U. Siddesh, C. N. R. Rao, Ferromagnetism as a universal feature of nanoparticles of the otherwise nonmagnetic oxides. Phys. Rev. B 74 (2006), 161306 (R). 4 pages.

[39]   S. J. Han, J. W. Song, C. H. Yang, S. H. Park, J. H. Park, Y. H. Jeong and K. W. Rhie, A key to room-temperature ferromagnetism in Fe-doped ZnO: Cu, Appl. Phys. Lett. 81 (2002), 4212. https://doi.org/10.1063/1.1525885

[40]   S. Dutta, M. Chakrabarti, D. Jana, D. Sanyal and A. Sarkar, Defect dynamics in annealed ZnO by positron annihilation spectroscopy, J. Appl. Phys. 98 (2005) 053513. https://doi.org/10.1063/1.2035308

[41]   F. Tuomisto, A. Mycielski and K. Grasza, Vacancy defects in (Zn, Mn)O, Superlattices and Microstructures 42 (2007) 218-221. https://doi.org/10.1016/j.spmi.2007.04.071

[42]   D. Sanyal, T. K. Roy, M. Chakrabarti, S. Dechoudhury, D. Bhowmick and A.
        Chakrabarti, Defect studies in annealed ZnO by positron annihilation
        spectroscopy, J. Phys., Condens. Matter 20 (2008), 045217.
        https://doi.org/10.1088/0953-8984/20/04/045217

[43]   A. Sarkar, H. Luitel, N. Gogurla and D Sanyal, Positron annihilation spectroscopic
        characterization of defects in wide band gap oxide semiconductors Mat. Sc. Exp. 4
        (2017), 35909.

[44]   D. C. Iza, D. Muñoz-Rojas , Q. Jia, B. Swartzentruber and J. L. Macmanus-
        Driscoll, Tuning of defects in ZnO nanorod arrays used in bulk heterojunction
        solar Cells, Nanoscale Res. Lett. 7 (2012), 655-662. https://doi.org/10.1186/1556-
        276X-7-655

[45]   A. Janotti and C. G. Van de Walle, Native point defects in ZnO. Phys. Rev. B 76
        (2007), 165202 (22pages)

[46]   H. Luitel,A. Sarkar, M. Chakrabarti, S. Chattopadhyay, K. Asokan and D. Sanyal
        Positron annihilation lifetime characterization of oxygen ion irradiated rutile
        TiO2 , Nuclear Instru.& Methods B 379 (2016), 215-218.
        https://doi.org/10.1016/j.nimb.2016.04.014

[47]   A. Sarkar, D. Sanyal, P. Nath, M. Chakrabarti, S. Pal, S. Chattopadhyay, D. Jana
        and K. Asokan, Defect driven ferromagnetism in SnO$_2$: a combined study using
        density functional theory and positron annihilation spectroscopy RSC Advances 5
        (2015), 1148-1152. https://doi.org/10.1039/C4RA11658E

[48]   J. W. Rasmussen, E. Martinez, P. Louka, D. G. Wingett, Zinc oxide
        nanoparticles for selective destruction of tumor cells and potential for drug
        delivery applications, Expert Opin Drug Deliv 7 (2010), 1063-1077.
        https://doi.org/10.1517/17425247.2010.502560

[49]   Ihab M. Obaidat, Borhan A. Albiss and Yousef Haik, Magnetic Nanoparticles:
        Surface Effects and Properties Related to Biomedicine Applications, Int J Mol
        Sci.14 (2013) 21266–21305. https://doi.org/10.3390/ijms141121266

Bio-Medical Applications

Nanomaterial scaffold holds the status as one of the critical research endeavors of the early 21st century, as scientists harness the unique properties of atomic and molecular assemblages built at the nanometer scale. Our ability to manipulate the physical, chemical, and biological properties of these particles affords researchers the capability to rationally design and use nanoparticles for drug delivery, as image contrast agents, and for diagnostic as well as therapeutic purposes. The confluence of these newly acquired capabilities, coupled with advances in imaging, bioinformatics, and systems biology, holds tremendous promise for answering some of biology's most challenging problems today.

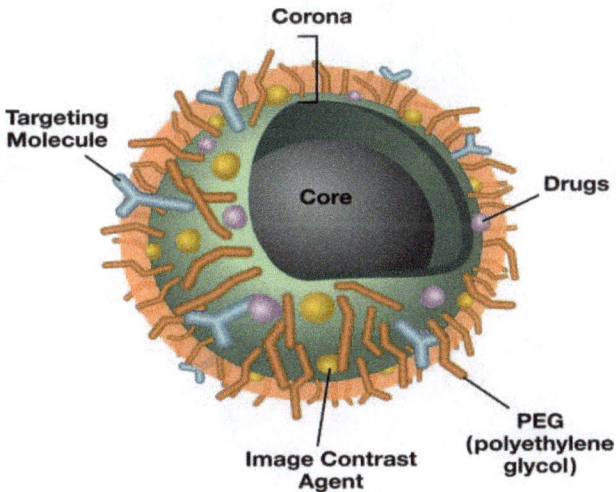

Chapter 7

Advances in the Application of Nanomaterials and Nanosacled Materials in Physiology or Medicine: Now and the Future

Uttam Pal[1] and Sumit Kumar Pramanik[2]

[1]Chemical Sciences Division, Saha Institute of Nuclear Physics, 1/AF Salt Lake, Kolkata 700064, India

[2]CSIR — Central Salt and Marine Chemicals Research Institute, Shree Gijubhai Badheka Rd, Bhavnagar, Gujarat 364002, India

uttam.pal@saha.ac.in (UP), sumitpramanik@csmcri.res.in (SKP)

"There's plenty of room at the bottom." -- Richard Feynman at American Physical Society meeting at Caltech on December 29, 1959.

Abstract

Nanomedicine is a booming field, however, the use of nanomaterials in medicine can be traced back thousands of years. Starting its journey in the dark ages of the gold ash in Indian traditional medicine through the Feynman's futuristic concept of swallowing surgeon, the nanomedicine recently entered into the era of exponential growth. Due to the reduction in size, nanomaterials provide a unique advantage for application in biology. Like building a magic bullet for treating cancer or detecting disease, the use of nanoparticles is everywhere. In this chapter, we have discussed the current exciting developments in this field and what the future holds.

Keywords

Anticancer, Phototherapy, Sensing, Imaging, Drug Delivery, Swallowing Surgeons

Contents

1. Introduction...148

2. Cytotoxic nanomaterials 150

    2.1   Anti-cancer agents 151

2.2   Anti-microbial and anti-fungal agents 152

3.   **Nontoxic nanomaterials** **152**

3.1   Carriers of drugs 153

3.2   Conjugated with nano 153

3.2.1 Porous materials 154

3.2.2 Capsules and liposomes 157

3.2.3 Adjuvants for vaccines 159

3.3   Phototherapeutic agents 159

3.4   Tissue engineering 161

4.   **Application in biomedical assays** **162**

4.1   Imaging 163

4.2   Sensing 164

4.3   Blood purification 166

5.   **Towards swallowing the surgeon** **167**

7.   **References** **170**

1.   Introduction

The possibility of tiny machines that can voyage through the blood into the remotely accessible regions of a human body to diagnose or repair problems with higher precision and less pain, eluded scientists as well as the fiction writers for decades. In his famous speech from 1959, "There's Plenty of Room at the Bottom", Nobel laureate physicist Richard Feynman, called this concept "swallow the surgeon". However, "the bottom" remains unexplored until the early nineties and then it exploded. Over the last few decades, the advances in materials science expanded its horizons into the boundaries of medical science. Various nanomaterials found their application in several aspects of therapeutics and the nano surgeon appears to be within reach. Now, nanomaterials are being used to treat cancer, for engineering tissue as well as to better detect pathological conditions. In this chapter, we shall discuss categorically the applications of several nanomaterials in the biomedical field based on their properties such as toxicity, composition, shape, photophysical behavior, etc. We shall also discuss how we can exploit those properties of a nanomaterial to fabricate their application in therapeutics.

The area of nanomedicine ranges from medical applications of nanomaterials and biological devices, to nanoelectronic biosensors, and even possible future applications of molecular nanotechnology such as biological machines. Current problems for nanomedicine involve understanding the issues related to toxicity and environmental impact of nanoscale materials. Functionalities can be added to nanomaterials by interfacing them with biological molecules or structures. The size of nanomaterials is similar to that of most biological molecules and structures; therefore, nanomaterials can be useful for both *in vivo* and *in vitro* biomedical research and applications. Thus, far along the path, the integration of nanomaterials with biology has led to the development of diagnostic devices, contrast agents, analytical tools, physical therapy applications, and drug delivery vehicles.

Indian traditional medicine [1] has a history of using nanomaterials in its formulations [2]. Herbo-metallic preparations such as *bhasmas* (ash) are used traditionally in Indian and Chinese medicinal systems. In Ayurveda, *Swarna* (gold) *bhasma* is used to treat several clinical manifestations. The major constituent of this gold ash has been found to be nanosized gold particles in the size range of 50-200 nm prepared by green synthesis [3]. Charaka Samhita (1500 BC) provides the details for reducing the particle size of metals (such as gold or iron). *Bhasmas* are, therefore, claimed to be green synthesized nanoparticles. Preparations containing metal ash are known in the Indian subcontinent since the seventh century AD and widely prescribed with several other medicines of Ayurveda for treatment of a variety of ailments.

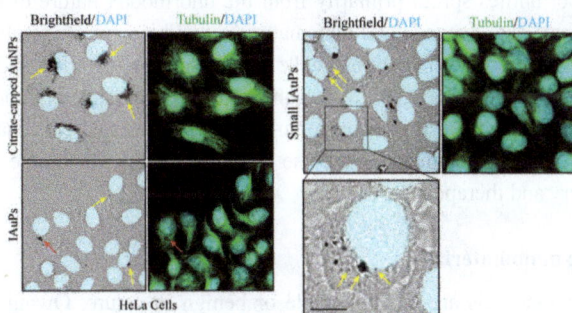

Figure 1: Cellular uptake and subcellular localization of citrated capped gold nanoparticles (AuNPs) of modern days and medieval gold ash (IAuPs). Reproduced from Beaudet et al. [3]. License: CC-BY.

149

In recent times, nano-sized materials have been explored to develop new life-saving technologies. Nanoparticles are used to make miniaturized sensors for nucleic acids, pathogens, and chemicals [4,5]. Ultra-strong polymer nanocomposites have been developed by using nanotubes, nanoclays, and ceramic nanoparticles, which finds their application in orthopedic medicine for example in joint-replacement. Nanoparticles are also used to make nano-textured substrates to support attachment and growth of cells for tissue engineering. Hydroxyapatite nanopowders, incorporated in biodegradable polymer composites [6] or deposited on biocompatible substrates [7], have been demonstrated to promote adhesion and proliferation of bone-forming cells.

High-quality nanomaterials of well-controlled size and shape provide a new class of building blocks to the development of assays for monitoring molecular signals in biological systems and living organisms. Many of these new nano-assays have been claimed to have higher sensitivity and selectivity than conventional bioanalytical methods and may be used in high throughput screening. Such nano-assays are capable of detecting biochemical changes at the single-molecule level in living cells [8]. Moreover, such assays will lead to the production of low-cost, point-of-care devices for rapid diagnosis of infections and genetic diseases (e.g., HIV and cancer) [9]. In addition, nanomaterials are already in use as advanced contrast agents for clinical imaging technologies, such as MRI, computer tomography and ultrasound diagnosis [10–12]. Further, the use of nanomaterials could lead to the invention of smart drug-delivery vehicles (magic bullets), new therapies and even new noninvasive life saving methods.

These new opportunities sprout primarily from the unorthodox nature of nanomaterials. Pertaining to their size-dependence, nanomaterials exhibit new physical and chemical properties compared with conventional bulk and molecular materials [13]. In general, nanomaterials include inorganic, organic and inorganic/organic composite nanostructures, such as nanoparticles, nanowires and nanopatterns. This chapter discusses the current exciting developments in the use of nanomaterials for biomedical diagnosis and drug delivery and therapeutics.

2.   Cytotoxic nanomaterials

Not all the nanomaterials are biocompatible or benign in nature. Owing to their small size, which is smaller than the cells and cellular organelles, nanomaterials may have the ability to penetrate the cell membrane and alter redox balance or disrupt the normal function of organelles. However, selective toxicity towards some cells or microorganisms, in turn, renders them useful in biomedical applications. Selective toxicity of the nanomaterials can be utilized to kill the microorganisms, pathogens or unwanted growth such as cancer. In this section, we shall discuss the direct application of cytotoxic nanoparticles as anticancer agents and anti-microbial agents. We shall also highlight the tinkering with these cytotoxic effects of particles so that they can be used for other purposes such as drug delivery vehicles. However, regardless of the cytotoxicity, any nanomaterial can be used *in vitro* biomedical assays, which shall be discussed later in this chapter.

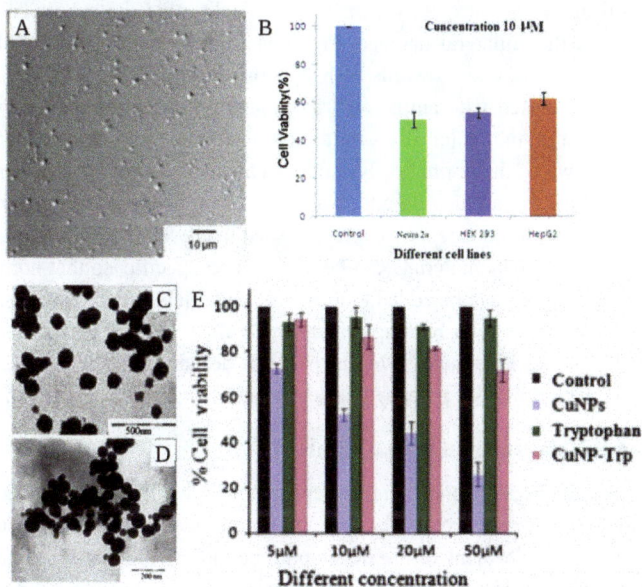

Figure 2: Scanning electron micrograph of a peptide nanoparticle (A) and its cytotoxicity towards different cancer cell lines (B). Reproduced from Banerji et al., 2012 [14]. Transmission electron microscopic images of Cu(I) nanoparticles (C) and its tryptophan conjugate (D). (E) Changes in the cytotoxicity due to surface modification, Reproduced from Maity et al., 2014 [15].

2.1   Anti-cancer agents

Some nanomaterials made of copper or other organic nanomaterials, even which are made of peptides, often show cytotoxicity. However, when their level of toxicity is more towards the cancer cells than the normal cells, they can be used to kill the cancer cells. We have shown previously, that a tetra peptide constellated with alternate D- and L-proline forms spherical nanostructures in solution (Figure 2A) [14] and shows no harmful effects toward cells. Interestingly, it was found to activate caspase 3 mediated apoptosis in three different types of cancer cells (Neura 2a, HEK 293 and Hep G2). Therefore, selectivity becomes an issue in their applications to kill cancer cells. Sometimes, the nanomaterials that show nonspecific toxicity can be coated with some other materials and such surface modification may reduce or alter their toxicity towards some cell types. Surface modification may also lead them to specifically bind the cancer cells and thus reduce the extent the collateral damage. Previous studies from our group showed that copper nanoparticles are toxic towards both the normal and cancer cells [15]. But when the materials were coated with amino acid tryptophan, those became less toxic towards the normal cells. With the realm of a similar approach many toxic nanomaterials can be made less toxic towards the normal cells and thus can be used as anticancer agents. Some small molecule ligands can also be added to the surface of these nanomaterials, which would go and bind with the receptors expressed specifically on the cancer cells. Thus, the delivery of these toxic nanomaterials can be made target specific, so that normal cells do not get affected. Markers can be added to these nanomaterials which recognizes receptors on the cancer cell, those upon binding initiate internalization of the nanomaterial inside the cancer cell and thus kills them. Over the past few decades various nanomaterials have been developed which showed potency to selective targeting of the cancer cells.

2.2   Anti-microbial and anti-fungal agents

Apart from the pathological condition that develops from within such as cancer, there are several conditions that arise from infections by microorganisms. Protozoa, parasites and fungal infections are prevalent in the developing countries. Malaria, leishmaniasis, etc. are caused by protozoa [16,17]. Fungal infection in the eye may blind a person. Many skin infections are caused by parasites such as ringworms are contagious. Nanoparticles that can kill such microorganisms can be used to treat such infections. In such cases, nanomaterials can also be used in combination with drugs to improve their potency. Micro-robots decorated with silver nanoparticles kill bacteria in aqueous media [18] as also antimicrobial gold nanoclusters [19]. Such nanoformulations are mostly used externally and maybe harmful in systemic administration.

3.   Nontoxic nanomaterials

As of now we have discussed the toxic nanomaterials and how they can be used to kill the external pathogens or cancerous cells. However, there are a plethora of nanomaterials that are not toxic to the cellular organisms. Most of the nanomaterials made of noble metals such as gold or silver show no or negligible toxicity to cells or bio-organisms.

Sometimes zinc based nanomaterials (above a certain range, can be non-toxic but not always) which are used in cosmetics and ointments are also non-toxic. Silica based nanomaterials are also nontoxic. Organic nanomaterials made of proteins or small peptides often show very little toxicity or immune response and therefore can be administered without severe consequences. Due to their benign behavior, these kinds of nanomaterials might not have direct applications in killing cancer cells or parasites. However, their non-toxic nature makes them very good carriers of drugs or materials for the synthesis of tissue scaffold on which cells can proliferate and make organs for transplantation. Such materials are also used in drug formulations for the controlled release of the drugs.

3.1 Carriers of drugs

A major focus of the material science nowadays is centered on the development of drug carriers. Small molecule drugs often show alternate targets. Due to this non-specificity, some drugs produce toxic side effects. If there were some vehicles that can carry the drug molecule to its site of action, it could minimize the toxic effect of the drug molecules. Most of the chemotherapeutic drugs are very toxic and take a toll on the body. Scientists are trying to develop carriers for such drugs that will deliver these small molecules to their desired site of action i.e. cancer sites. By this way, the bioavailability and distribution of the drug could be controlled precisely. Moreover, it will reduce the dose to a minimum. There could be two possible ways to use nanomaterials as drug carriers. One way would be to covalently attach the drug molecules to the surface of the nanoparticles. The other method would be to load the drug into the cage of hollow or porous nanomaterials. Further, many modifications can be done to facilitate pH or reactive oxygen species dependent release. Such environment sensitive non-toxic nanomaterials shall act as a magic bullet to treat cancer and various other localized infections (such as tuberculosis) as well.

3.1.1 Conjugated with nano

Gold nanoparticles are widely used to deliver drugs. Gold and silver nanoparticles are generally nontoxic and, therefore, suitable for this purpose. Drugs can be adsorbed on the gold nanoparticle surface or may be attached covalently through thiol or anime modifications. Based on the modifications drugs can be released at controlled pH or in presence of reactive oxygen species. The efficiency of drug loading and release can be measured spectroscopically. Recently from our group, Sau et al. showed that, riboflavin in conjugation with gold nanoparticles may translocate into the cellular nucleus of cancer cells and induce apoptosis via DNA damage [20,21]. Riboflavin is a vitamin and does not generally enter into the nucleus, however, if it can be translocated into the nucleus, it has the capability to intercalate with DNA and undergoes covalent modification of guanine base pair resulting in irreversible DNA damage. Furthermore, cancer cells show markers for riboflavin more than the normal cells thus the riboflavin conjugated nanoparticle can specifically target the cancer cells.

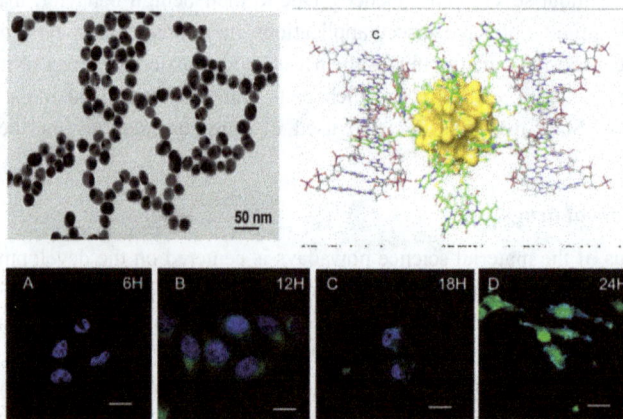

Figure 3: Riboflavin coated gold nanoparticles (top left) and theoretical model of riboflavin coated gold nanoparticle mediated DNA damage (top right). Bottom panel shows the real time confocal images of treated HeLa cells at different time points. Scale Bar represents 20 μm. Reproduced from Sau et al., 2018 [20].

3.1.2 Porous materials

Porous nanoparticles such as mesoporous silica offer one of the best possible drug delivery system through the release of drug molecules from the accessible pores. Often mesoporous silica is preferred due to their biocompatibility, high surface area, controllable pore size, cost-effective easy synthetic routes and capability of drug molecule loading and releasing. Porous nanoparticles are capable of encapsulating functional molecules into their pores to protect them from the physiological environment and their known chemical inertness make them a suitable drug delivery system for effective chemotherapy. For these properties, porous materials such as mesoporous silica nanoparticles have been extensively tested for drug and gene delivery applications.

In a model drug delivery study by Bardhan et al. on squamous cells showed that the amount of drug delivered to the cells by silica nanoparticles could be much higher compared to direct administration (Figure 4) [22]. The implication of this could be in the effective reduction of the ED50 (effective dose) of a drug molecule with the help of silica nanoparticles. ED50 value determines how much of the drug is required to be administered per kg of body weight of an individual. When the ED50 of a drug is low, less amount of the substance is required to be administered. A nano formulation such as with mesoporous silica can often improve the cellular availability of drugs many folds, thus, reducing the ED50 of a drug by orders of magnitude. Therefore, costly life saving drugs could be made affordable to the common mass if such nano-vehicles are used.

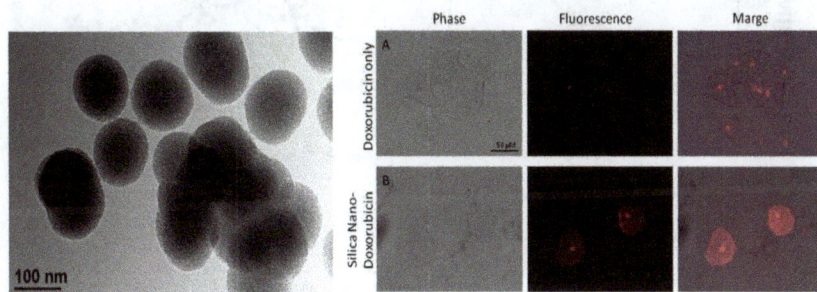

Figure 4: A TEM image of drug loaded mesoporous silica nanoparticles (left) and comparison of its drug delivery potential to live cells (right panel). Reproduced from Bardhan et al., 2018 [22].

In a separate study, Bhandary et al. demonstrated a pH responsive mesoporous silica-based nanoplatform for delivery of chemotherapeutic drugs [23]. Unlike the previous study, they have functionalized the mesoporous silica nanoparticles to achieve this pH responsive property. Schematics of the functionalization is shown in figure 5. At first, they functionalized the mesoporous silica with amines, which was subsequently modified with 4-carboxyphenylboronic acid for anchoring β-cyclodextrin (β-CD) onto the nanoparticles with a pH responsive boronic acid-catechol ester bond that will break at pH 5.0. Afterward, drug molecules were encapsulated into the hydrophobic core of the mesopores of silica. PEGylation and folic acid incorporation were performed by mPEG functionalized with adamantine (mPEG-Ada), and Ada unit functionalized with the foliate unit (FA-Ada). Folic acid has been used to target the vehicle specifically towards the cancer cells as certain cancers cells are known to overexpress folic acid receptors whereas PEGylation makes it more soluble, biocompatible and suppresses non-specific protein adsorption of the nanoformulations. Finally, β-CD was added through complexation with Ada, which has been used as gatekeepers to control the release of drug molecules from the pores of the nanoparticles. Fluorescent molecules can also be added to this vehicle through the amine linkage, which would enable visualization of cellular uptake.

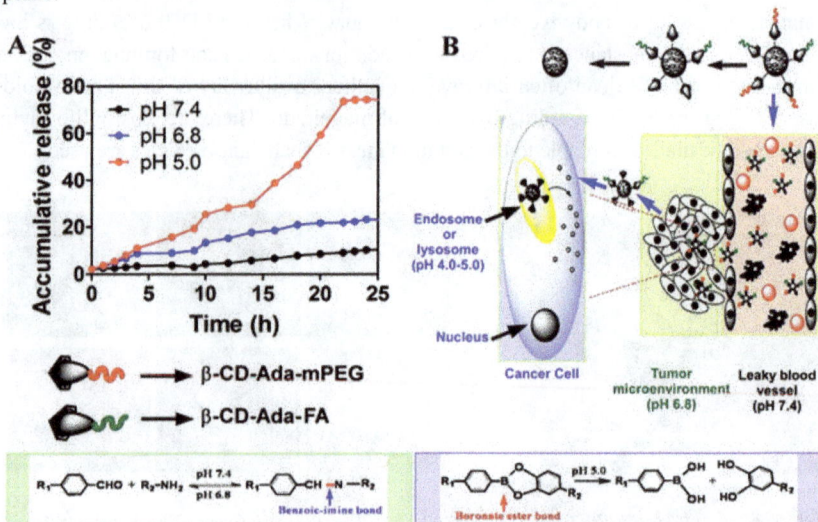

Figure 5: pH responsive drug release from a functionalized mesoporous silica based nano vehicle (A). Schematics of the pH response cascade (B). Reproduced from Bhandary et al., 2017 [23].

This nanoplatform was shown to be very efficient in delivery and targeting with respect to the treatment with drug molecule alone [23,24]. The concept of this nanomachinery worked because it made use of pH-sensitive benzoic-imine linkages that remains stable at physiological pH because of the proper π-π conjugation. Under weak acidic tumor extracellular (pH ~6.8), these linkages begin partially hydrolyzing to accelerate mPEG release and to facilitate internalization of the whole complex. Further, the boronic acid-catecholester bonds that hold the gatekeepers (Figure 5) were stable at pH 7.4 and a relatively small fraction of cleavage of these linkers occurres in the tumor microenvironment However, when exposed to endosomal pH, most of the linkers are cleaved, leading to β-CD dissociation, unblocking of the nanotunnels and eventually release of the drug molecules into the cytoplasm. Thus, such pH-responsive mesoporous nonmaterial could achieve on demand release of anticancer drugs exhibiting its wide applicability and even great potential for tumor therapy in response to cascade pH stimuli within the tumor microenvironment.

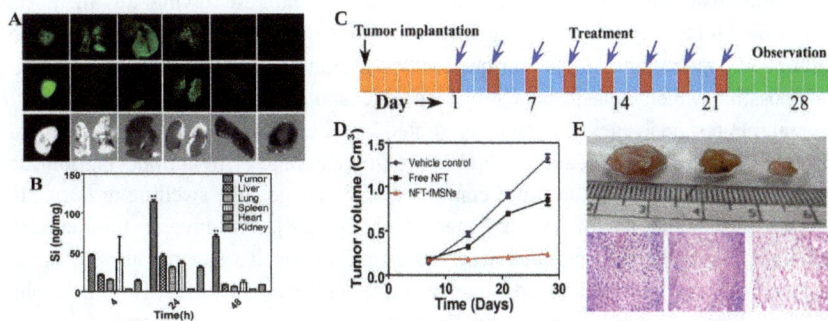

Figure 6: Ex vivo fluorescence imaging (A) of excised tumor, lung, liver, kidney, spleen and heart isolated from fluorescence labeled and functionalized mesoporous silica nanoparticle treated mice after 4 h (upper panel) and 24 h (middle panel) of treatment. Lower panel shows white light images. (B) Organ accumulation of Si at 4 h, 24 h and 48 h. (C) Cancer induction and treatment regime with drug (NFT) loaded nanoparticles (NFT-fMSNs). Animals received injections (red boxes) on every 3rd day after the first injection for 22 days. (D) Tumor growth curves. (E) comparative photograph of the collected tumor tissues at day 28 from vehicle control (left side), free NFT treated (middle) and NFT-fMSNs treated (right side) groups and their respective H&E stained sections (lower panel) showing significant tumor inhibitory effect of NFT-fMSNs.Reproduced from Bhandary et al., 2017 [23].

3.1.3 Capsules and liposomes

Peptides have the unique ability to self-assemble to various nanostructures [14,25,26]. By modifying the peptides, the shape and size of the oligomeric assembly can be altered and fine-tuned. Such peptide based materials can be used in drug delivery systems. Peptides are bio-molecules and therefore often show biocompatibility and tend to be nontoxic. However, sometimes they can produce an immune response and often get degraded by proteases present in the blood. These limitations of the peptide based nanomaterials can be overcome by using amino acids or side chain modification and using non-canonical peptide bonds [14]. Peptide capsules formed via disulfide linkages can be easily broken by changing the pH or in presence of radicals or reactive oxygen species. Thus, they can be good environmental sensors as well, making them suitable for targeted drug delivery. Apart from the peptides, other nanomaterials such as synthetic polymers and liposomes etc can also be used to encapsulate drugs. Here we shall discuss two such platforms: polymers and liposomes.

Polymeric nanocapsules have attracted a lot of interest owing to their high biocompatibility, very low immunogenicity and the ability for the encapsulation of both hydrophobic/hydrophilic drugs. Furthermore, post grafting of the nanocapsule surface is also possible which makes them target specific and tailored degradation *in vivo*. In general, pH responsiveness can be incorporated into the polymeric nanocarriers either by incorporating linkages that are prone to hydrolytic cleavage or by introducing ionizable groups (acidic/basic) that can cause conformational changes like swelling or deswelling or enhenced solubility of the polymer chains [27,28]. An inverse mini-emulsion technique can be used to produce drug loaded pH responsive nanocapsules containing polyurethane polythiourethane and biodegradable polyester linkages. The nucleophilic addition of thiols to isocyanates formed thiourethane (–NH–CO–S–) linkages whereas the nucleophilic addition of hydroxyls to isocyanates yielded the urethane (–NH–CO–O–) linkages [29]. The aqueous core nanocapsules are synthesized via an interfacial polyaddition reaction of thiol terminated Poly(L-lactide) and toluene isocyanate. The reaction takes place at the interface of inverse-miniemulsion droplets and the polythiourethane-polyurethane-polylactide nanocapsules are obtained with the drug encapsulated as a hydrophilic payload.

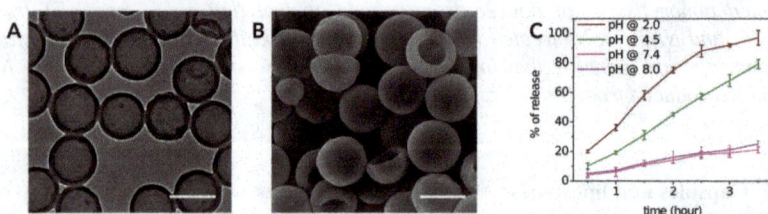

Figure 7: (A) TEM (scale bar 150 nm) and (B) SEM (scale bar 150 nm) images of a pH responsive polymeric nanocapsule. (C) Release of drug at different pH. Pramanik et al. (unpublished data).

The other system that we shall discuss is the liposome platform. These are nano scaled materials and may not be strictly called nanomaterials. However, liposomal technology has secured a fortified position as a potential nanocarrier. Recently, we have reported the physicochemical characterizations of functional hybrid liposomal nanocarriers formed using photo-sensitive lipids [30]. Specifically, radiation/photo-sensitive liposomes containing photo-polymerizable cross-linking lipids are intriguing as they can impart the vesicles with highly interesting properties such as response to stimulus and improved shell stability. Photo-polymerizable lipid such as 1,2-bis(10,12-tricosadiynoyl)-sn-glycero-3-phosphoethanolamine (DTPE) may be used to form functional hybrid-liposomes as it can form intermolecular cross-linking through the diacetylene groups. Usage of mixtures containing polymerizable lipids when designing new systems has potential applications as nanocarriers for new drug delivery system. Tinkering with the trigger responsiveness of the photosensitive liposomes and the incorporation of lipids with suitable functional end groups for conjugation of (bio-)molecules can lead to the development of advanced nanocarriers with useful functionalities enabling target specific release of therapeutics for in vivo drug delivery applications.

3.2 Adjuvants for vaccines

Sometimes, mare presence of a nanomaterial along with the active component may render higher efficacy. Adjuvants are such examples and nanomaterials can serve as excellent adjuvants. Vaccines offer the most cost-effective solution to prevent infectious and possibly non-infectious diseases. Nanoparticles are also shown to be used as an adjuvant for vaccines. An adjuvant is a pharmacological or immunological agent that modifies the effect of other agents. It has been shown to increase or prolong the antibody responses against a variety of immunogens by maximizing exposure to the immune system [31]. Some nanoparticles can both stabilize vaccine antigens and enter antigen-presenting cells by different pathways, thereby modulating the immune response to the antigen, which may be critical for the induction of immune responses to intracellular pathogens [32]. The most commonly used adjuvants are aluminium based but these can induce local reactions and may fail to generate strong cell-mediated immunity. Only two of them: a group of aluminium derivatives including aluminium phosphate, aluminium hydroxyphosphate, and aluminium hydroxide (Alum) and monophosphoryl lipid A (MPL) are licensed by the FDA for use in human vaccines today [33]. As the adjuvant is critical for the augmentation of the proper immunogenicity of the biological component, which obviously is not able to induce strong humoral and cellular immune responses [34]. Skrastina et al. demonstrated a silica nanoparticles as the adjuvant for the Immunisation of mice using Hepatitis B core virus-like particles [34].

3.3 Phototherapeutic agents

Some metallic nontoxic optically active nanomaterials can directly be used as phototherapeutic agents. It has been demonstrated that gold nanoparticles can be used in such photothermal therapy. Based on their size, gold nanoparticles exerts various spectroscopic behaviors. They absorb IR radiation and produce heat. Therefore, such properties can be utilized to kill cancer cells or other targets. Gold nanoparticles can be easily uptaken by the cells. Gold deposits inside the cells can be observed in transmission electron microscope (Figure 1). Therefore, upon administration of gold nanoparticles near tumor site, if IR light is radiated, gold particles inside the cells shall absorb the IR frequencies and produce heat which in turn fires the cancer cells. Other nontoxic metallic nanomaterials can also be used in such photothermal therapy. Figure 9 describes an application of a gold based plasmonic nano-heater in photothermal therapy by Jørgensen et al. [35].

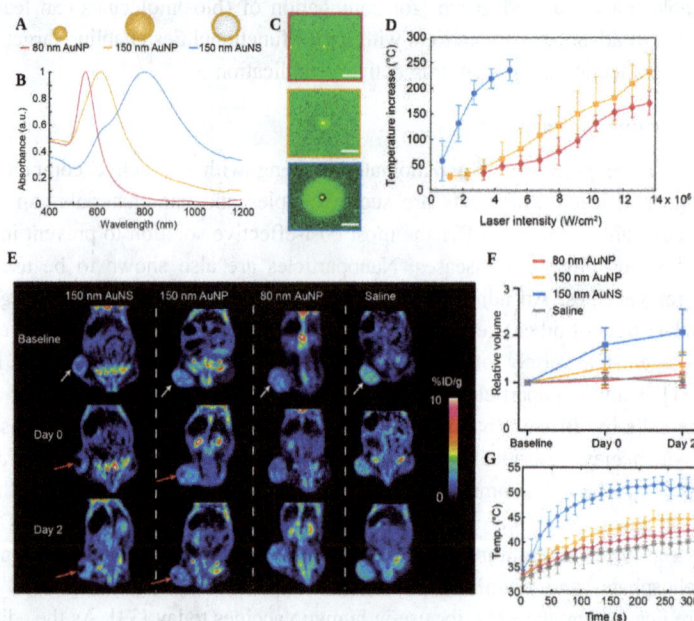

Figure 8: Sketch of the investigated plasmonic nanoparticles (A) and their absorption spectra in water (B). (C) Local melting due to optically heating a nanoparticle embedded in the bilayer with a tightly focused laser beam. Scale bars are 2 µm. (D) The surface temperature as a function of laser intensity for all three nanoparticles. (E) ^{18}F-D-glucose PET scan of mice from each nanoparticle group (columns) before, immediately after (Day 0), and two days post-treatment (Day 2). Tumor location is marked with white arrow, where the nanoparticles were administered locally and subsequently irradiated with the laser. AuNS and 150 nm AuNP treated groups show a subvolume (marked with red arrows) with decreased uptake of ^{18}F-D-glucose both at Day 0 and Day 2. (F) The mean relative changes in the tumor volume with low ^{18}F-D-glucose uptake in the four treatment groups. (G) The temperature evolution at the skin surface of the tumors of mice. Reproduced from Jørgensen et al., 2016 [35]. License: CC-BY.

The transduction of photons by specific molecules/materials into other types of energy such as heat or chemical energy forms the basis for the development of photothermal/photochemical therapies (collectively called photodynamic therapy). Nanomaterials are designed such that they absorb near infrared (NIR) wavelengths, particularly in the range of ≈650–1450 nm, which is especially advantageous since this range falls in the so called "optically transparent window," and light can penetrate relatively deeply into biological materials as there is minimal photon absorption by the tissue [36–38]. Such nano-constructs may enable diagnosis and treatment of diseases at high spatial resolution. In photodynamic therapy, light may also be used to produce high energy oxygen molecules which will chemically react with and destroy most organic molecules that are next to them. Such therapies are appealing for many reasons as they do not leave a toxic trail of reactive molecules throughout the body. It is restricted to the region where only the light is focused and the particles exist. It is potentially a noninvasive procedure for dealing with infections, growth and tumors.

3.4   Tissue engineering

Nanomaterials found its way into making scaffolds for tissue engineering. In tissue engineering, the major aim is to help reproduce or repair or reshape damaged tissue using suitable nanomaterial-based scaffolds and growth factors. Tissue engineering if successful may replace conventional treatments like organ transplants or artificial implants. Two-dimensional materials based on single-walled and multiwalled carbon nanotubes and nanoparticles crosslinked with biodegradable and biocompatible polymers provide excellent building blocks for scaffolds on which cells and adhere and grow to form tissue [39]. Mechanically strong polymer composites made of ceramic nanoparticles and nanoclays showed potential applications in orthopedic medicine such as bone implants. However, here we shall focus briefly on the self-assembled peptide based nanomaterials. There has been a rapid advancement in the development of self-assembled nano-biomaterials, such as nanotubes, nanospheres, and nanowires, which have potential applications in generating scaffolds for tissue engineering. Tunable nature of the physical and chemical properties of these nanomaterials by controlling their shapes and sizes are very attractive features for research. Polymers or peptides are routinely used to design various scaffolds. There are various designing rules of the synthesis of these biomaterials, following which the secondary structure of the self-assembled fibers, their thickness, porosity, and other different mechanical properties can all be varied very predictably [40–43]. Over the years, various natural self-assembling systems have served as inspiration for the design of novel building blocks on the nanoscale. It has been already reported that cyclic peptides, amphiphilic peptides, and amyloid-inspired peptides can form ordered nanostructures with different morphologies including nanowires, nanotubes, nanovesicles, nanofibrils, and nanosheets that can further organize to form nanoscale scaffolds.

We have precisely reported a Cysteine based dipeptide that could undergo self-assembly and form unbranched hollow nanotubes in aqueous solution [25]. Phenylalanine was already known to form such nanotubes, however, it was one of the first reports for cysteine-based nanotubes. Pure cysteine dipeptide does not form tubular structure but when the sulfide groups are benzylated, which promotes pi-stacking interactions, ordered self-assembly follows (Figure 9). When we further tweaked this peptide to make it a tripeptide, it interestingly resulted in an annular protofibrillar assembly instead of a tubular structure [26]. It demonstrates the tunability at the molecular level for the design of peptide based self-assembly building blocks with potential applications in tissue engineering.

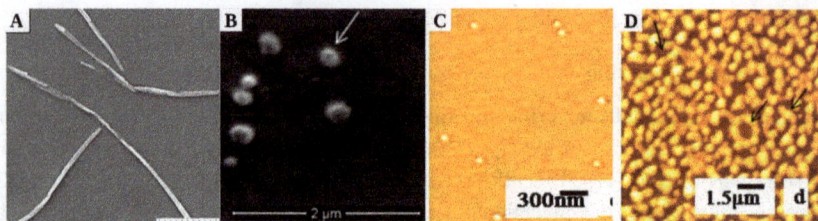

Figure 9: (A) Self assembled peptide nanotubes obtained from benzyl protected cysteine dipeptide in aqueous solution. Scale bar 10 μm. Reproduced from Banerji et al., 2013 [25]. (B) Tripeptide produced dough-nut shaped assembly of ~300 nm diameter in ethylacetate solution, which breaks down to ~70 nm spheres (C) upon sonication. (D) Higher concentration of the tripeptide produced higher order assembly structures. Reproduced from Banerji et al., 2017 [26].

4.   Application in biomedical assays

Apart from creating direct interference with the human physiology, nanoparticles found a niche in the development of imaging, sensing, etc. to aid health-care systems. For example, magnetic core shell nanoparticles have shown very promising results in detecting circulating cancer cells, which could not be detected efficiently by any other method. Nanoparticles also can serve as contrast agents in imaging. In this section, applications of nanostructures in biodiagnostics [9] shall be addressed. However, it should be noted that there is no clear demarcation between diagnostics and therapeutics when it comes to developing nanoplatforms for biomedical applications. Nanomaterials are often designed to serve these dual-purpose and thus called theranostics.

4.1 Imaging

Nanoparticles be it metallic or organic often respond to light stimuli. Metallic nanoparticles may show plasmonic properties based on their sizes. On the other hand, carbon-based nanoparticles are also generally inherently fluorescent. Graphine based particles show excitation independent fluorescence behavior, allowing us to choose excitation wavelength to reduce background emission. Metallic quantum dots show visible emission, whereas larger particles may also absorb in NIR and upconvert it into visible light before emission. These diverse properties of nanoparticles make them an excellent medium for imaging. Recently we have reported a lysosome targeted imaging agent, a surface modified fluorescence upconversion nanoparticle for real time probing of this organelle's role in cellular processes (Figure 10) [44]. Such nanoplatform is superior to the conventional molecular fluorescence probes as the latter suffers from photobleaching and cytotoxicity. Besides, molecular probes that emit in visible range requires UV irradiation, which causes further damage and nonspecific emission from internal fluorophores, whereas, upconversion nanoparticles can be excited with low energy NIR light from the optically transparent window.

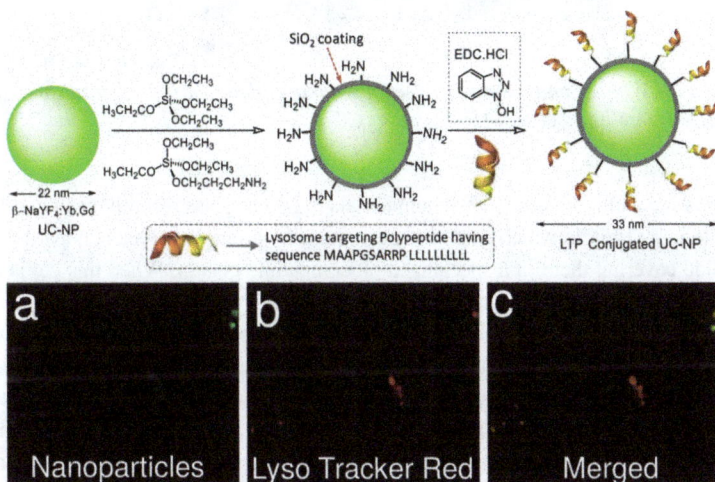

Figure 10: Upconversion nano particles for imaging. A schematic representation for the development of a lysosome targeting peptide conjugated upconversion naoparticles (top panel). Comparison of the images obtained by applying nanoparticles with the conventional small molecule, lysotracker dye. Reproduced from Pramanik et al., 2017 [44].

We have discussed photodynamic therapy before [35]. When such nanoparticles are conjugated with fluorescence probes, they can act as phototheranostic agents [45]. Figure 11 shows this combined phototheranostic activity of gold nanorods conjugated with porphyrin as fluoresce probe for imaging and an antibody for targeting [45]. In fact, any drug delivery vehicles may be conjugated with a fluorescent probe to turn them into trackers. Sometimes the cargo itself may be tracked by visual stimulus. The mesoporous silica based nanomaterials [22,23] or the liposomes [30] that had been discussed earlier, all have the added advantage of imaging capability to them.

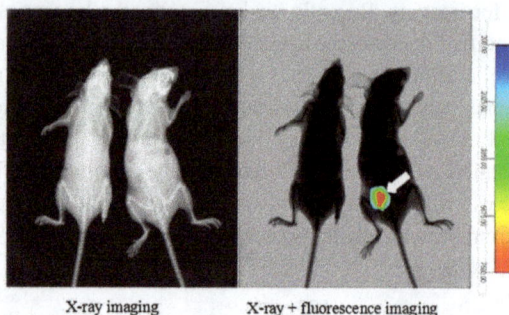

X-ray imaging   X-ray + fluorescence imaging

Figure 11: Whole body imaging using nanoparticle based theranostic agents. Reproduced from Kang et al. [45] (License: CC-BY).

An interesting and indirect use of nanomaterials in imaging has recently been highlighted in the literature [46], in which gels have been made by arranging silica nanoparticles on a glass slide covering them in a specialized biocompatible gelatin. When this gel is peeled off from the silica layer, it leaves an imprint of tiny nanoparticles on the gel, which makes it display vivid structural colors. Researchers have grown heart cells in these gels, which when contract, transfer the motion to the gels resulting in the shift in structural colors in synchrony of the pulsating heart cells. A virtual heart on chip could be developed from it by combining other materials, which may be used for biomedical testing instead of using human or animal organs. It shows no direct use of nanoparticles, but only the imprint of it was used to develop a biomedical assay, which enables drug testing using only the naked eye.

4.2   Sensing

Sensors are the molecules or materials that reports about their milieu by sending visual or some other detectable cues [47,48]. Fluorescent nanoparticles and quantum dots can, therefore, be used as sensors. Thus, sensing by nanomaterials adds one more dimension to the lab-on-a-chip technology. Examples of nanomaterial-based sensing include the use of magnetic or plasmonic nanoparticles bound to a suitable antibody that can label specific molecules, structures or microorganisms; gold nanoparticles tagged with short segments of DNA for detection of genetic sequence in a sample; multicolor optical coding by quantum dot polymer composites and so on. The previously described upconversion nanoparticle [44] is also a sensor for lysosome. Porphyrin and antibody tagged gold nanorods described by Kang et al. for photothermal therapy can specifically sense the targeted cancer cells [45]. Apart from the metallic nanoparticles, other polymeric or carbon based materials can also be used for sensing. Bera et al. recently reported a fluorescent carbon dot that can image cells as well as report the presence of toxic quinone compounds in real time [49].

Electrochemical sensing with nanomaterials is another method with applications *in vitro* assays. Metallic nanoparticles often show catalytic activity, the extent of which depends on the types of crystal defects they harbor. Oxidation reduction potential of such a nanoparticle may be exploited to produce electrocatalytic sensors. Using these nanoparticles as electrode materials and measuring differential pulse voltammetric spectra, the biomarkers may be detected in pathological samples [50,51]. The peak potential differs significantly for different closely related substances (such as ascorbic acid, uric acid, dopamine, glucose) present in the test sample. Therefore, using this method simultaneous multicomponent sensing in blood samples may be performed efficiently. Further, the stability and reusability of nanomaterials make the technology very much cost effective.

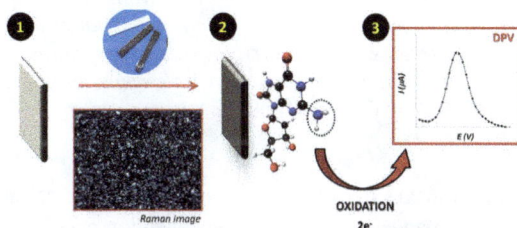

Figure 12: Schematics of electrocatalytic sensing of an oxidative stress biomarker 8-Hydroxy-2'-deoxyguanosine on a conductive carbon paper substrate. Hydrophobic white paper as substrate (1), conductive carbon-coating (2) followed by in-situ electrochemical (differential pulse voltammetric) measurement (3). Reproduced from Martins et al., 2017 [50]. License: CC-BY.

4.3 Blood purification

One of the direct ways to prevent or cure infection is by removing the unwanted molecules from blood or body fluids. Dialysis has long been used in medical science to remove small molecules like urea to filter out in patients with renal dysfunction. However, this method is not very selective and works on the principle of the size related diffusion of solutes and ultrafiltration of fluid across a semi-permeable membrane. In contrast to dialysis, the purification with nanoparticles allows specific targeting of substances. Additionally, larger compounds which are commonly not dialyzable can be removed.

Magnetic nanoparticles hold immense potential for the separation of cells and proteins from complex media. Magnetic-activated cell sorting is one of such available technology. More recently it was shown in animal models that magnetic nanoparticles can be used for the removal of various noxious compounds including toxins, pathogens, and proteins from whole blood in an extracorporeal circuit similar to dialysis [52,53].

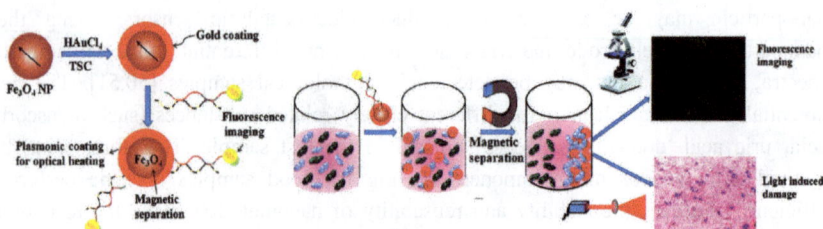

Figure 13: Schematic representation showing the synthesis of aptamer-conjugated multifunctional magnetic core—gold shell nanoparticles and their applications in separating of specific cancer cells followed by selective fluorescence imaging and targeted photothermal destruction. Reproduced with permission from Fan et al., 2012 [59]. Copyright 2012 American Chemical Society.

The purification process makes use of functionalized iron oxide or carbon coated metal nanoparticles with ferromagnetic or superparamagnetic properties [54]. The carbon coating prevents the iron core from being oxidized or losing magnetism. Incorporation of some other noble metal such as platinum makes them traceable in an iron rich environment. Binding agents such as proteins, antibodies, antibiotics, or synthetic ligands are covalently linked to the particle surface [52,53,55,56]. These binding agents are able to interact with target species forming an agglomerate. Applying an external magnetic field gradient allows exerting a force on the nanoparticles, thus, separating them from the bulk fluid, thereby cleaning it from the contaminants [57,58].

The small size and large surface area of functionalized nano-magnets lead to advantageous properties compared to hemoperfusion, which is a clinically used technique for the purification of blood and is based on surface adsorption. The advantages are high loading and accessibility of the binding agents, high selectivity towards the target compound, fast diffusion, small hydrodynamic resistance, and low dosage. Further, Fan et al. describe a plasmonic shell–magnetic core nanoparticle that might be used for targeted trapping, isolation and photothermal destruction of circulating metastatic tumor cells [59].

This approach offers new therapeutic possibilities for the treatment of systemic infections such as sepsis by directly removing the pathogen [53]. It can also be used to selectively remove cytokines or endotoxins [60] or for the dialysis of compounds which are not accessible by traditional dialysis methods. However, the technology is not yet available for clinical trials as a major challenge is to making sure that nanoparticles do not enter the bloodstream. Using ultra-strong magnetic nanoparticles may help to overcome this challenge.

5. Towards swallowing the surgeon

After its conception in 1959 and a long lag phase, development and fabrication of mechanical devices powered by artificial molecular machines have recently become one of the primary goals of nanoscience [61]. In recognition of nanotechnology's huge potential as an emerging technology field creating machines or robots whose components are at or near the scale of a nanometre, 2016 Nobel Prize in chemistry was awarded to Jean-Pierre Sauvage, Sir J. Fraser Stoddart and Bernard L. Feringa, for their work on developing molecules with controllable movements [62,63]. In the 1980s, Sauvage and Stoddart pioneered the syntheses of mechanically-interlocked molecular architectures such as catenanes and rotaxanes utilizing molecular recognition and molecular self-assembly [64–66]. Their work bridged the gap between chemistry and the scientific and engineering challenges of nanoelectromechanical systems. In the late 1990s, Feringa's work in stereochemistry led to major contributions in photochemistry, resulting in the first monodirectional light driven molecular rotary motor and later a molecular car driven by electrical impulses [67,68].

There are many types of nanorobots which have already been developed in the last two decades and are potentially applicable [69–73]. A detailed discussion is beyond the scope of this chapter. Some examples are discussed here. Scientists at Ohio State University have designed and constructed complex nanoscale mechanical parts using 'DNA origami' and can produce complex, controllable components for future nanorobots [74]. Nanoswimmers have been built by the scientists at ETH Zurich and Technion [75]. These are elastic polypyrrole (Ppy) nanowires of about 15 micrometers long and 200 nanometers thick that can move through biological fluid environments at almost 15 micrometers per second. The nanoswimmers could be functionalized to deliver drugs and be magnetically controlled to swim through the bloodstream to target cancer cells, for example. Self-propelled, synthetic active matters that can transduce chemical energy into mechanical motion are of great current interest [70]. Many of the physical challenges associated with generating motility on the micro- and nanoscale have recently been overcome [70]. Katuri et al. recently described self-propelled colloids which are the synthetic analogs of biological self-propelled units such as algae or bacteria and also powered by flagellar motion [76]. Li et al. fabricated magnetic nanoswimmers capable of efficient "freestyle" swimming [77]. Propulsion of the two-arm nanorobot is attributed to synchronized oscillatory deformations of the nanorobot under the combined action of the magnetic field and viscous forces. Improvements in speed regulation and remote navigation open up the possibilities in designing remotely actuated nanorobots for biomedical operation at the nanoscale [77]. Enzymes are now being tested for harnessing and converting free chemical energy into kinetic forces in order to power nanomachines [78,79]. Examples include high speed DNA based rolling motors powered by RNAse H [80]. An acoustic based propulsion system to maneuver nanosized objects is also under development. Ahmed et al. described a new class of nanoswimers propelled by small-amplitude oscillation produced by flagellum-like flexible tail [81]. There are several other motion manipulations such as ultrasound, electric field, temperature, pH, chemotaxis etc. [82,83].

Baumberg et al. from the Cavendish Laboratory at University of Cambridge researchers have developed a tiny engine called actuating nanotransducers (ANTs), which are capable to exert a force per unit-weight nearly 100 times higher than any motor or muscle [84]. The new nano-engines could lead to nanorobots small enough to enter living cells to fight diseases. Other examples include sperm-inspired microrobots [85]. Researchers at the University of Twente (Netherlands) and German University in Cairo (Egypt) have developed sperm-inspired microrobots, which can be controlled by oscillating weak magnetic fields. They will be used in complex micro-manipulation and targeted therapy tasks. Then there are bacteria-powered robots. Drexel University engineers have developed a method for using electric fields to help microscopic bacteria-powered robots detect obstacles in their environment and navigate around them [86,87]. Uses include delivering medication, manipulating stem cells to direct their growth, or building a microstructure, such as a self-propelled metal–polymer hybrid micromachines with bending and rotational motions [88].

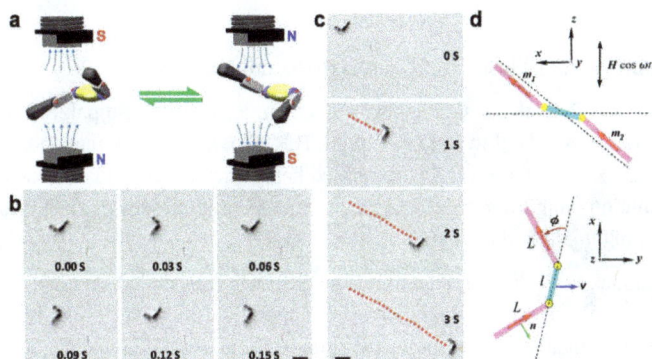

Figure 14: Freestyle propulsion of two-arm nanoswimmer with an oscillating magnetic field. (a) Schematic showing the magnetic setup for propulsion along with the vibrating magnetic field. (b)Time lapse images depicting the efficient freestyle propulsion of the nanoswimmer. Scale bar, 2 µm. (c) Tracking lines illustrates the travel distances of the freestyle nanoswimmer over 1s, 2s and 3s. Scale bar, 2 µm. (d) Dynamics of two-arm nanoswimmer and its shapes in different planes. Reproduced with permission from Li et al., 2017 [77]. Copyright 2017 American Chemical Society.

Peng et al. in their recent review highlighted the recent efforts towards realistic *in vivo* applications of various motor systems [71]. The first-generation microrobots, which could deliver therapeutics and other cargo to targeted specific body sites, have just been started to be tested in small animals for clinical use. Ceylan et al. in their compressive review frontline advances in design, fabrication, and testing of untethered mobile microrobots for bioengineering applications [72]. Theranostic applications of these nano and micro-machines are potentially endless.

6. Conclusion

This chapter reviews the various modes of applications of nanomaterials or nanoscale materials in physiology or medicine, yet it only scratched the surface. There has been an exponential growth of targeted research in nanomedicine, such that the interface of synthetic smart materials and biomedical science is overwhelming and there is no true systematic way to assimilate it. We addressed the subject here, first with relevance to toxicity and then the possible mode of applications. However, the multifunctionality and modularity of the nanoplatform are the features that make it a fascinating field to explore.

7. References

[1] A. Dance, Medical histories, Nature. (2016). doi:10.1038/537S52a.

[2] B.N. Singh, Prateeksha, G. Pandey, V. Jadaun, S. Singh, R. Bajpai, S. Nayaka, A.H. Naqvi, A.K.S. Rawat, D.K. Upreti, B.R. Singh, Development and characterization of a novel Swarna-based herbo-metallic colloidal nano-formulation – inhibitor of Streptococcus mutans quorum sensing, RSC Adv. 5 (2014) 5809–5822. doi:10.1039/C4RA11939H.

[3] D. Beaudet, S. Badilescu, K. Kuruvinashetti, A.S. Kashani, D. Jaunky, S. Ouellette, A. Piekny, M. Packirisamy, Comparative study on cellular entry of incinerated ancient gold particles (Swarna Bhasma) and chemically synthesized gold particles, Sci. Rep. 7 (2017) 10678. doi:10.1038/s41598-017-10872-3.

[4] P. Alonso-Cristobal, P. Vilela, A. El-Sagheer, E. Lopez-Cabarcos, T. Brown, O.L. Muskens, J. Rubio-Retama, A.G. Kanaras, Highly Sensitive DNA Sensor Based on Upconversion Nanoparticles and Graphene Oxide, ACS Appl. Mater. Interfaces. 7 (2015) 12422–12429. doi:10.1021/am507591u.

[5] T. Mocan, C.T. Matea, T. Pop, O. Mosteanu, A.D. Buzoianu, C. Puia, C. Iancu, L. Mocan, Development of nanoparticle-based optical sensors for pathogenic bacterial detection, J. Nanobiotechnology. 15 (2017) 25. doi:10.1186/s12951-017-0260-y.

[6] S.-S. Kim, M. Sun Park, O. Jeon, C. Yong Choi, B.-S. Kim, Poly(lactide-co-glycolide)/hydroxyapatite composite scaffolds for bone tissue engineering, Biomaterials. 27 (2006) 1399–1409. doi:10.1016/j.biomaterials.2005.08.016.

[7] M. Sato, M.A. Sambito, A. Aslani, N.M. Kalkhoran, E.B. Slamovich, T.J. Webster, Increased osteoblast functions on undoped and yttrium-doped nanocrystalline hydroxyapatite coatings on titanium, Biomaterials. 27 (2006) 2358–2369. doi:10.1016/j.biomaterials.2005.10.041.

[8] X. Michalet, F.F. Pinaud, L.A. Bentolila, J.M. Tsay, S. Doose, J.J. Li, G. Sundaresan, A.M. Wu, S.S. Gambhir, S. Weiss, Quantum dots for live cells, in vivo imaging, and diagnostics, Science. 307 (2005) 538–544. doi:10.1126/science.1104274.

[9]   N.L. Rosi, C.A. Mirkin, Nanostructures in Biodiagnostics, Chem. Rev. 105 (2005) 1547–1562. doi:10.1021/cr030067f.

[10]   J.-H. Lee, Y.-M. Huh, Y. Jun, J. Seo, J. Jang, H.-T. Song, S. Kim, E.-J. Cho, H.-G. Yoon, J.-S. Suh, J. Cheon, Artificially engineered magnetic nanoparticles for ultra-sensitive molecular imaging, Nat. Med. 13 (2006) nm1467. doi:10.1038/nm1467.

[11]   D. Kim, S. Park, J.H. Lee, Y.Y. Jeong, S. Jon, Antibiofouling Polymer-Coated Gold Nanoparticles as a Contrast Agent for in Vivo X-ray Computed Tomography Imaging, J. Am. Chem. Soc. 129 (2007) 7661–7665. doi:10.1021/ja071471p.

[12]   J.H. Sakamoto, B.R. Smith, B. Xie, S.I. Rokhlin, S.C. Lee, M. Ferrari, The molecular analysis of breast cancer utilizing targeted nanoparticle based ultrasound contrast agents, Technol. Cancer Res. Treat. 4 (2005) 627–636. doi:10.1177/153303460500400606.

[13]   M.G. Bawendi, M.L. Steigerwald, L.E. Brus, The Quantum Mechanics of Larger Semiconductor Clusters ("Quantum Dots"), Annu. Rev. Phys. Chem. 41 (1990) 477–496. doi:10.1146/annurev.pc.41.100190.002401.

[14]   B. Banerji, S.K. Pramanik, U. Pal, N.C. Maiti, Conformation and cytotoxicity of a tetrapeptide constellated with alternative D- and L-proline, RSC Adv. 2 (2012) 6744–6747. doi:10.1039/C2RA20616A.

[15]   M. Maity, S.K. Pramanik, U. Pal, B. Banerji, N.C. Maiti, Copper(I) oxide nanoparticle and tryptophan as its biological conjugate: a modulation of cytotoxic effects, J. Nanoparticle Res. 16 (2014) 2179. doi:10.1007/s11051-013-2179-z.

[16]   A. Alam, S. Haldar, H.V. Thulasiram, R. Kumar, M. Goyal, M.S. Iqbal, C. Pal, S. Dey, S. Bindu, S. Sarkar, U. Pal, N.C. Maiti, U. Bandyopadhyay, Novel anti-inflammatory activity of epoxyazadiradione against macrophage migration inhibitory factor: inhibition of tautomerase and proinflammatory activities of macrophage migration inhibitory factor, J. Biol. Chem. 287 (2012) 24844–24861. doi:10.1074/jbc.M112.341321.

[17]   S. Saha, C. Acharya, U. Pal, S.R. Chowdhury, K. Sarkar, N.C. Maiti, P. Jaisankar, H.K. Majumder, A novel spirooxindole derivative inhibits the growth of Leishmania donovani parasite both in vitro and in vivo by targeting type IB topoisomerase, Antimicrob. Agents Chemother. (2016) AAC.00352-16. doi:10.1128/AAC.00352-16.

[18] D. Vilela, M.M. Stanton, J. Parmar, S. Sánchez, Microbots Decorated with Silver Nanoparticles Kill Bacteria in Aqueous Media, ACS Appl. Mater. Interfaces. 9 (2017) 22093–22100. doi:10.1021/acsami.7b03006.

[19] K. Zheng, M.I. Setyawati, D.T. Leong, J. Xie, Antimicrobial Gold Nanoclusters, ACS Nano. 11 (2017) 6904–6910. doi:10.1021/acsnano.7b02035.

[20] A. Sau, S. Sanyal, K. Bera, S. Sen, A.K. Mitra, U. Pal, P.K. Chakraborty, S. Ganguly, B. Satpati, C. Das, S. Basu, DNA Damage and Apoptosis Induction in Cancer Cells by Chemically Engineered Thiolated Riboflavin Gold Nanoassembly, ACS Appl. Mater. Interfaces. (2018). doi:10.1021/acsami.7b18837.

[21] A. Sau, S. Sen, K. Bera, U. Pal, B. Satpati, C. Das, S. Basu, Nuclear Uptake of Thiolated Riboflavin Gold Nanoassembly: DNA Damage and Apoptosis Induction in Cancer Cell, Biophys. J. 114 (2018) 360a. doi:10.1016/j.bpj.2017.11.2002.

[22] M. Bardhan, A. Majumdar, S. Jana, T. Ghosh, U. Pal, S. Swarnakar, D. Senapati, Mesoporous silica for drug delivery: Interactions with model fluorescent lipid vesicles and live cells, J. Photochem. Photobiol. B. 178 (2018) 19–26. doi:10.1016/j.jphotobiol.2017.10.023.

[23] P.C.S. Suman Bhandary Arijit Bhowmik, Aparajita Ghosh, Suchandrima Saha, Uttam Pal, Nivedita Roy, Nilanjan Chakraborty, Arijit Chakraborty, Mrinal K. Ghosh, Targeting IL-6/IL-6R Signaling Axis in Triple-Negative Breast Cancer by a Novel Nifetepimine-Loaded Cascade pH Responsive Mesoporous Silica Based Nanoplatform, Glob. J. Nanomedicine. 3 (2017) 555609. doi:10.19080/GJN.2017.03.555609.

[24] S. Ghosh, A. Adhikary, S. Chakraborty, P. Bhattacharjee, M. Mazumder, S. Putatunda, M. Gorain, A. Chakraborty, G.C. Kundu, T. Das, P.C. Sen, Cross-talk between Endoplasmic Reticulum (ER) Stress and the MEK/ERK Pathway Potentiates Apoptosis in Human Triple Negative Breast Carcinoma Cells ROLE OF A DIHYDROPYRIMIDONE, NIFETEPIMINE, J. Biol. Chem. 290 (2015) 3936–3949. doi:10.1074/jbc.M114.594028.

[25] B. Banerji, S.K. Pramanik, U. Pal, N.C. Maiti, Dipeptide derived from benzylcystine forms unbranched nanotubes in aqueous solution, J. Nanostructure Chem. 3 (2013) 12. doi:10.1186/2193-8865-3-12.

[26] B. Banerji, M. Chatterjee, U. Pal, N.C. Maiti, Formation of Annular Protofibrillar Assembly by Cysteine Tripeptide: Unraveling the Interactions with NMR, FTIR,

and Molecular Dynamics, J. Phys. Chem. B. 121 (2017) 6367–6379. doi:10.1021/acs.jpcb.7b04373.

[27]   S. Mura, J. Nicolas, P. Couvreur, Stimuli-responsive nanocarriers for drug delivery, Nat. Mater. 12 (2013) 991–1003. doi:10.1038/nmat3776.

[28]   N. Kamaly, B. Yameen, J. Wu, O.C. Farokhzad, Degradable Controlled-Release Polymers and Polymeric Nanoparticles: Mechanisms of Controlling Drug Release, Chem. Rev. 116 (2016) 2602–2663. doi:10.1021/acs.chemrev.5b00346.

[29]   E. Delebecq, J.-P. Pascault, B. Boutevin, F. Ganachaud, On the Versatility of Urethane/Urea Bonds: Reversibility, Blocked Isocyanate, and Non-isocyanate Polyurethane, Chem. Rev. 113 (2013) 80–118. doi:10.1021/cr300195n.

[30]   S.K. Pramanik, P. Losada-Pérez, G. Reekmans, R. Carleer, M. D'Olieslaeger, D. Vanderzande, P. Adriaensens, A. Ethirajan, Physicochemical characterizations of functional hybrid liposomal nanocarriers formed using photo-sensitive lipids, Sci. Rep. 7 (2017) 46257. doi:10.1038/srep46257.

[31]   J. Kreuter, Nanoparticles as adjuvants for vaccines, Pharm. Biotechnol. 6 (1995) 463–472.

[32]   A.E. Gregory, R. Titball, D. Williamson, Vaccine delivery using nanoparticles, Front. Cell. Infect. Microbiol. 3 (2013). doi:10.3389/fcimb.2013.00013.

[33]   R. Rappuoli, C.W. Mandl, S. Black, E. De Gregorio, Vaccines for the twenty-first century society, Nat. Rev. Immunol. 11 (2011) 865–872. doi:10.1038/nri3085.

[34]   D. Skrastina, I. Petrovskis, I. Lieknina, J. Bogans, R. Renhofa, V. Ose, A. Dishlers, Y. Dekhtyar, P. Pumpens, Silica Nanoparticles as the Adjuvant for the Immunisation of Mice Using Hepatitis B Core Virus-Like Particles, PLoS ONE. 9 (2014) e114006. doi:10.1371/journal.pone.0114006.

[35]   J.T. Jørgensen, K. Norregaard, P. Tian, P.M. Bendix, A. Kjaer, L.B. Oddershede, Single Particle and PET-based Platform for Identifying Optimal Plasmonic Nano-Heaters for Photothermal Cancer Therapy, Sci. Rep. 6 (2016) 30076. doi:10.1038/srep30076.

[36]   B. Bahmani, D. Bacon, B. Anvari, Erythrocyte-derived photo-theranostic agents: hybrid nano-vesicles containing indocyanine green for near infrared imaging and therapeutic applications, Sci. Rep. 3 (2013) 2180. doi:10.1038/srep02180.

[37]   V. Pansare, S. Hejazi, W. Faenza, R.K. Prud'homme, Review of Long-Wavelength Optical and NIR Imaging Materials: Contrast Agents, Fluorophores

and Multifunctional Nano Carriers, Chem. Mater. Publ. Am. Chem. Soc. 24 (2012) 812–827. doi:10.1021/cm2028367.

[38]   N. Goswami, Z. Luo, X. Yuan, D.T. Leong, J. Xie, Engineering gold-based radiosensitizers for cancer radiotherapy, Mater. Horiz. 4 (2017) 817–831. doi:10.1039/C7MH00451F.

[39]   G. Lalwani, A.M. Henslee, B. Farshid, L. Lin, F.K. Kasper, Y.-X. Qin, A.G. Mikos, B. Sitharaman, Two-Dimensional Nanostructure-Reinforced Biodegradable Polymeric Nanocomposites for Bone Tissue Engineering, Biomacromolecules. 14 (2013) 900–909. doi:10.1021/bm301995s.

[40]   E. Gazit, Self-assembled peptide nanostructures: the design of molecular building blocks and their technological utilization, Chem. Soc. Rev. 36 (2007) 1263–1269. doi:10.1039/B605536M.

[41]   N. Stephanopoulos, J.H. Ortony, S.I. Stupp, Self-Assembly for the Synthesis of Functional Biomaterials, Acta Mater. 61 (2013) 912–930. doi:10.1016/j.actamat.2012.10.046.

[42]   G. Fichman, E. Gazit, Self-assembly of short peptides to form hydrogels: design of building blocks, physical properties and technological applications, Acta Biomater. 10 (2014) 1671–1682. doi:10.1016/j.actbio.2013.08.013.

[43]   N. Annabi, J.W. Nichol, X. Zhong, C. Ji, S. Koshy, A. Khademhosseini, F. Dehghani, Controlling the Porosity and Microarchitecture of Hydrogels for Tissue Engineering, Tissue Eng. Part B Rev. 16 (2010) 371–383. doi:10.1089/ten.teb.2009.0639.

[44]   S.K. Pramanik, S. Sreedharan, H. Singh, N.H. Green, C. Smythe, J.A. Thomas, A. Das, Imaging cellular trafficking processes in real time using lysosome targeted up-conversion nanoparticles, Chem. Commun. (2017). doi:10.1039/C7CC08185E.

[45]   X. Kang, X. Guo, X. Niu, W. An, S. Li, Z. Liu, Y. Yang, N. Wang, Q. Jiang, C. Yan, H. Wang, Q. Zhang, Photothermal therapeutic application of gold nanorods-porphyrin-trastuzumab complexes in HER2-positive breast cancer, Sci. Rep. 7 (2017) 42069. doi:10.1038/srep42069.

[46]   F. Fu, L. Shang, Z. Chen, Y. Yu, Y. Zhao, Bioinspired living structural color hydrogels, Sci. Robot. 3 (2018) eaar8580. doi:10.1126/scirobotics.aar8580.

[47]   U. Pal, S.K. Pramanik, B. Bhattacharya, B. Banerji, N.C. Maiti, Binding interaction of a novel fluorophore with serum albumins: steady state fluorescence

perturbation and molecular modeling analysis, SpringerPlus. 4 (2015) 548. doi:10.1186/s40064-015-1333-8.

[48] U. Pal, Interaction of proteins with small molecules and peptides, Doctoral dissertation, Jadavpur University, 2016. http://www.eprints.iicb.res.in/id/eprint/2550 (accessed September 9, 2016).

[49] K. Bera, A. Sau, P. Mondal, R. Mukherjee, D. Mookherjee, A. Metya, A.K. Kundu, D. Mandal, B. Satpati, O. Chakrabarti, S. Basu, Metamorphosis of Ruthenium-Doped Carbon Dots: In Search of the Origin of Photoluminescence and Beyond, Chem. Mater. 28 (2016) 7404–7413. doi:10.1021/acs.chemmater.6b03008.

[50] G.V. Martins, A.P.M. Tavares, E. Fortunato, M.G.F. Sales, Paper-Based Sensing Device for Electrochemical Detection of Oxidative Stress Biomarker 8-Hydroxy-2′-deoxyguanosine (8-OHdG) in Point-of-Care, Sci. Rep. 7 (2017) 14558. doi:10.1038/s41598-017-14878-9.

[51] Singh K., Solanki Pratima R., Basu Tinku, Malhotra B. D., Polypyrrole/multiwalled carbon nanotubes-based biosensor for cholesterol estimation, Polym. Adv. Technol. 23 (2011) 1084–1091. doi:10.1002/pat.2020.

[52] I.K. Herrmann, A. Schlegel, R. Graf, C.M. Schumacher, N. Senn, M. Hasler, S. Gschwind, A.-M. Hirt, D. Günther, P.-A. Clavien, W.J. Stark, B. Beck-Schimmer, Nanomagnet-based removal of lead and digoxin from living rats, Nanoscale. 5 (2013) 8718–8723. doi:10.1039/C3NR02468G.

[53] J.H. Kang, M. Super, C.W. Yung, R.M. Cooper, K. Domansky, A.R. Graveline, T. Mammoto, J.B. Berthet, H. Tobin, M.J. Cartwright, A.L. Watters, M. Rottman, A. Waterhouse, A. Mammoto, N. Gamini, M.J. Rodas, A. Kole, A. Jiang, T.M. Valentin, A. Diaz, K. Takahashi, D.E. Ingber, An extracorporeal blood-cleansing device for sepsis therapy, Nat. Med. 20 (2014) 1211. doi:10.1038/nm.3640.

[54] C.C. Berry, A.S.G. Curtis, Functionalisation of magnetic nanoparticles for applications in biomedicine, J. Phys. Appl. Phys. 36 (2003) R198. doi:10.1088/0022-3727/36/13/203.

[55] J.-J. Lee, K.J. Jeong, M. Hashimoto, A.H. Kwon, A. Rwei, S.A. Shankarappa, J.H. Tsui, D.S. Kohane, Synthetic Ligand-Coated Magnetic Nanoparticles for Microfluidic Bacterial Separation from Blood, Nano Lett. 14 (2014) 1–5. doi:10.1021/nl3047305.

[56]   I.K. Herrmann, M. Urner, S. Graf, C.M. Schumacher, B. Roth-Z'graggen, M. Hasler, W.J. Stark, B. Beck-Schimmer, Endotoxin Removal by Magnetic Separation-Based Blood Purification, Adv. Healthc. Mater. 2 (2013) 829–835. doi:10.1002/adhm.201200358.

[57]   C.M. Schumacher, I.K. Herrmann, S.B. Bubenhofer, S. Gschwind, A.-M. Hirt, B. Beck-Schimmer, D. Günther, W.J. Stark, Quantitative Recovery of Magnetic Nanoparticles from Flowing Blood: Trace Analysis and the Role of Magnetization, Adv. Funct. Mater. 23 (2013) 4888–4896. doi:10.1002/adfm.201300696.

[58]   C.W. Yung, J. Fiering, A.J. Mueller, D.E. Ingber, Micromagnetic–microfluidic blood cleansing device, Lab. Chip. 9 (2009) 1171–1177. doi:10.1039/B816986A.

[59]   Z. Fan, M. Shelton, A.K. Singh, D. Senapati, S.A. Khan, P.C. Ray, Multifunctional Plasmonic Shell–Magnetic Core Nanoparticles for Targeted Diagnostics, Isolation, and Photothermal Destruction of Tumor Cells, ACS Nano. 6 (2012) 1065–1073. doi:10.1021/nn2045246.

[60]   E.J. Kwon, J.H. Lo, S.N. Bhatia, Smart nanosystems: Bio-inspired technologies that interact with the host environment, Proc. Natl. Acad. Sci. U. S. A. 112 (2015) 14460–14466. doi:10.1073/pnas.1508522112.

[61]   A. Coskun, M. Banaszak, R. Dean Astumian, J. Fraser Stoddart, B. A. Grzybowski, Great expectations: can artificial molecular machines deliver on their promise?, Chem. Soc. Rev. 41 (2012) 19–30. doi:10.1039/C1CS15262A.

[62]   R. Van Noorden, D. Castelvecchi, World's tiniest machines win chemistry Nobel, Nat. News. 538 (2016) 152. doi:10.1038/nature.2016.20734.

[63]   J.C. Barnes, C.A. Mirkin, Profile of Jean-Pierre Sauvage, Sir J. Fraser Stoddart, and Bernard L. Feringa, 2016 Nobel Laureates in Chemistry, Proc. Natl. Acad. Sci. 114 (2017) 620–625. doi:10.1073/pnas.1619330114.

[64]   C.O. Dietrich-Buchecker, J.P. Sauvage, J.P. Kintzinger, Une nouvelle famille de molecules : les metallo-catenanes, Tetrahedron Lett. 24 (1983) 5095–5098. doi:10.1016/S0040-4039(00)94050-4.

[65]   P.R. Ashton, T.T. Goodnow, A.E. Kaifer, M.V. Reddington, A.M.Z. Slawin, N. Spencer, J.F. Stoddart, C. Vicent, D.J. Williams, A [2] Catenane Made to Order, Angew. Chem. Int. Ed. Engl. 28 (1989) 1396–1399. doi:10.1002/anie.198913961.

[66]   R.A. Bissell, E. Córdova, A.E. Kaifer, J.F. Stoddart, A chemically and electrochemically switchable molecular shuttle, Nature. 369 (1994) 369133a0. doi:10.1038/369133a0.

[67]   N. Koumura, R.W.J. Zijlstra, R.A. van Delden, N. Harada, B.L. Feringa, Light-driven monodirectional molecular rotor, Nature. 401 (1999) 43646. doi:10.1038/43646.

[68]   T. Kudernac, N. Ruangsupapichat, M. Parschau, B. Maciá, N. Katsonis, S.R. Harutyunyan, K.-H. Ernst, B.L. Feringa, Electrically driven directional motion of a four-wheeled molecule on a metal surface, Nature. 479 (2011) nature10587. doi:10.1038/nature10587.

[69]   H. Wang, M. Pumera, Fabrication of Micro/Nanoscale Motors, Chem. Rev. 115 (2015) 8704–8735. doi:10.1021/acs.chemrev.5b00047.

[70]   K.K. Dey, A. Sen, Chemically Propelled Molecules and Machines, J. Am. Chem. Soc. 139 (2017) 7666–7676. doi:10.1021/jacs.7b02347.

[71]   F. Peng, Y. Tu, D. A. Wilson, Micro/nanomotors towards in vivo application: cell, tissue and biofluid, Chem. Soc. Rev. 46 (2017) 5289–5310. doi:10.1039/C6CS00885B.

[72]   H. Ceylan, J. Giltinan, K. Kozielski, M. Sitti, Mobile microrobots for bioengineering applications, Lab. Chip. 17 (2017) 1705–1724. doi:10.1039/C7LC00064B.

[73]   X.-Z. Chen, M. Hoop, F. Mushtaq, E. Siringil, C. Hu, B.J. Nelson, S. Pané, Recent developments in magnetically driven micro- and nanorobots, Appl. Mater. Today. 9 (2017) 37–48. doi:10.1016/j.apmt.2017.04.006.

[74]   A.E. Marras, L. Zhou, H.-J. Su, C.E. Castro, Programmable motion of DNA origami mechanisms, Proc. Natl. Acad. Sci. 112 (2015) 713–718. doi:10.1073/pnas.1408869112.

[75]   B. Jang, E. Gutman, N. Stucki, B.F. Seitz, P.D. Wendel-García, T. Newton, J. Pokki, O. Ergeneman, S. Pané, Y. Or, B.J. Nelson, Undulatory Locomotion of Magnetic Multilink Nanoswimmers, Nano Lett. 15 (2015) 4829–4833. doi:10.1021/acs.nanolett.5b01981.

[76]   J. Katuri, X. Ma, M.M. Stanton, S. Sánchez, Designing Micro- and Nanoswimmers for Specific Applications, Acc. Chem. Res. 50 (2017) 2–11. doi:10.1021/acs.accounts.6b00386.

[77]   T. Li, J. Li, K.I. Morozov, Z. Wu, T. Xu, I. Rozen, A.M. Leshansky, L. Li, J. Wang, Highly Efficient Freestyle Magnetic Nanoswimmer, Nano Lett. 17 (2017) 5092–5098. doi:10.1021/acs.nanolett.7b02383.

[78]   X. Ma, A.C. Hortelão, T. Patiño, S. Sánchez, Enzyme Catalysis To Power
       Micro/Nanomachines, ACS Nano. 10 (2016) 9111–9122.
       doi:10.1021/acsnano.6b04108.

[79]   E. Del Grosso, A.-M. Dallaire, A. Vallée-Bélisle, F. Ricci, Enzyme-Operated
       DNA-Based Nanodevices, Nano Lett. 15 (2015) 8407–8411.
       doi:10.1021/acs.nanolett.5b04566.

[80]   K. Yehl, A. Mugler, S. Vivek, Y. Liu, Y. Zhang, M. Fan, E.R. Weeks, K. Salaita,
       High-speed DNA-based rolling motors powered by RNase H, Nat. Nanotechnol.
       11 (2015) nnano.2015.259. doi:10.1038/nnano.2015.259.

[81]   D. Ahmed, T. Baasch, B. Jang, S. Pane, J. Dual, B.J. Nelson, Artificial Swimmers
       Propelled by Acoustically Activated Flagella, Nano Lett. 16 (2016) 4968–4974.
       doi:10.1021/acs.nanolett.6b01601.

[82]   W.Z. Teo, M. Pumera, Motion Control of Micro-/Nanomotors, Chem. – Eur. J. 22
       (2016) 14796–14804. doi:10.1002/chem.201602241.

[83]   Y. Tu, F. Peng, D.A. Wilson, Motion Manipulation of Micro- and Nanomotors,
       Adv. Mater. 29 (2017) n/a-n/a. doi:10.1002/adma.201701970.

[84]   T. Ding, V.K. Valev, A.R. Salmon, C.J. Forman, S.K. Smoukov, O.A. Scherman,
       D. Frenkel, J.J. Baumberg, Light-induced actuating nanotransducers, Proc. Natl.
       Acad. Sci. 113 (2016) 5503–5507. doi:10.1073/pnas.1524209113.

[85]   I.S.M. Khalil, H.C. Dijkslag, L. Abelmann, S. Misra, MagnetoSperm: A
       microrobot that navigates using weak magnetic fields, Appl. Phys. Lett. 104
       (2014) 223701. doi:10.1063/1.4880035.

[86]   H. Kim, U.K. Cheang, M.J. Kim, K. Lee, Obstacle avoidance method for
       microbiorobots using electric field control, in: 4th Annu. IEEE Int. Conf. Cyber
       Technol. Autom. Control Intell., 2014: pp. 117–122.
       doi:10.1109/CYBER.2014.6917446.

[87]   H. Kim, M.J. Kim, Electric Field Control of Bacteria-Powered Microrobots Using
       a Static Obstacle Avoidance Algorithm, IEEE Trans. Robot. 32 (2016) 125–137.
       doi:10.1109/TRO.2015.2504370.

[88]   Y. Yoshizumi, H. Suzuki, Self-Propelled Metal–Polymer Hybrid Micromachines
       with Bending and Rotational Motions, ACS Appl. Mater. Interfaces. 9 (2017)
       21355–21361. doi:10.1021/acsami.7b03656.

Chapter 8

Applications of Nanoparticles in Biomedicine

Bichitra Nandi Ganguly

Saha Institue of Nuclear Physics, KOLKATA-700064, India

Email:bichitra.ganguly@saha.ac.in

Abstract

Understanding of interactions between nanoparticles and bio-systems is essential for the effective utilization of these materials in biomedicine. A wide variety of nanoparticle surface structures have been developed for imaging, sensing, and drug delivery applications. In this research highlight, advances in tailoring nanoparticle interfaces for implementation in nanomedicine have been discussed. Nanoparticles exhibit unique physical properties such that when their size range commensurate with bio-molecular and cellular systems, their features make them attractive materials for therapeutic and diagnostic applications. Specifically designed nanoparticle monolayer structures can impart enhanced cellular internalization ability, noncytotoxicity and improved payload binding capacity necessary for effective intracellular delivery. Similarly, surface functionality can be tuned to provide the selective or specific recognition required for bio-sensing. Tailoring particle interfaces is a challenging task, chemists have a well-equipped toolbox to provide functionality through synthesis. Using different strategies, nanoparticles have been functionalized with a variety of ligands such as small molecules, surfactants, dendrimers, polymers, and bio-molecules. Biomolecule-conjugated nanoparticles can impart desired properties such as specific recognition or biocompatibility. The ease of such surface conjugation allows material scientists to create the desired functionalities for their future application in clinics. In this article different categories of nanoparticles *viz.* metal oxide, plasmonic metallic nanoparticles, surface coroneted nanoparticles etc. and their involvement in bio-application have been discussed.

Keywords

Biomedicine, Therapeutic and Diagnostic Applications, Surface Functionality, Drug-Delivery

Contents

1.   Introduction..180

2.   Nanomaterial physical properties and biological applications.........182

3.   Use of Metal Oxide Nanoparticles in Early Cancer Detection..........184

4.   Biomedical application of plasmonic nanoparticles :.......................186

5.   Immunoassay with nanoparticles...188

6.   Diagnostic study of tumor by radioactive imaging.............................189

7.   Conclusion 191

References...192

1.   Introduction

The recent trend in creating nanoparticles (nano engineering/nanotechnology) has opened up a new regime in material science and very optimistic challenges for their application in bioscience in general, especially in medical applications [1]. High quality nanomaterials of controlled size and shape are a new class of building blocks to enable establishment of assays for monitoring molecular signals in biological systems and living organisms.

Nanotechnology represents a new and enabling platform that promises to provide a broad range of novel uses and improved technologies for biological and biomedical applications. One of the reasons behind the intense interest is that nanotechnology permits the controlled synthesis of materials where at least one dimension of the structure is less than 100 nm. This ultra-small size is comparable to naturally occurring proteins and biomolecules in the cell, and is notably smaller than the typical diameter (~7 μm) of many human cells. The reduction of materials to the nanoscale can frequently alter their electrical, magnetic, structural, morphological, and chemical properties enabling them to interact in unique ways with cell biomolecules and enable their physical transport into the interior structures of cells. Nanoscale particles typically possess a larger percentage of atoms at the material's surface, which can lead to increased surface reactivity, and can maximize their ability to be loaded with therapeutic agents to deliver them to target cells. By appropriate engineering design, these nanomaterials can acquire the ability to selectively target particular types of cells or to pass through physiological barriers and penetrate deep into tumor sites which are exemplified in this article in the subsequent paragraphs [2].

Using well-established concepts and methods, while assessing the uniqueness of biomedical questions, nanoparticles have been used to develop ultrasensitive probes for HIV infection and cancer, among others diseases [3]. Novel directions using nanoparticles include theranostics and plasmonic photothermal therapy (PPT) [4], whereas understanding the fate of the nanoparticles once they are administered *in vivo* is a crucial aspect that is under thorough investigation [5]. In addition nanomaterials have been used as advanced contrast agents for clinical imaging technologies such as MRI, computer tomography[6,7]. Core concepts of materials science that have led to novel and exciting applications in biomedicine are highlighted in this chapter.

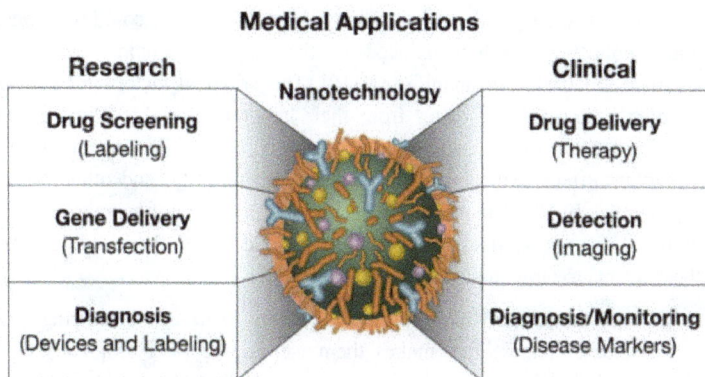

Medical Applications

Research	Nanotechnology	Clinical
Drug Screening (Labeling)		**Drug Delivery** (Therapy)
Gene Delivery (Transfection)		**Detection** (Imaging)
Diagnosis (Devices and Labeling)		**Diagnosis/Monitoring** (Disease Markers)

Figure 1. Medical applications of nanomaterials. The size and tailorability of nanoparticles may lead to their widespread use in a variety of medical applications.(ref 2.and copyright: J. Leukoc. Biol. 78: 585–594; 2005)

The application of nanotechnology to medical applications, commonly referred to as "nanomedicine", seeks to deliver a new set of tools, devices and therapies for treatment of human disease. Nanomaterials that can act as biological mimetics, "nanomachines", biomaterials for tissue engineering, shape-memory polymers as molecular switches, biosensors, laboratory diagnostics, and nanoscale devices for drug release, are just a few of the applications currently being explored [8].

As such, there is considerable interest in the role of nanomaterials for the rational delivery and targeting of pharmaceutical and diagnostics agents for the treatment of dreaded cancer. The potential use of ultra small nanoparticles of metallic cluster and other metal oxide nanoparticles in biomedical and cancer applications is gaining interest in the scientific and medical communities, largely due to the physical and chemical properties of these nanomaterials, and is the focus of this article.

2.   Nanomaterial physical properties and biological applications

The interfacing of material science study and biology provides the opportunity for the development of a new domain in the nanometer size range that can be applied to many potential applications in clinical medicine [2,9]. The most widely studied type of nanomaterials is the nanoparticle, which is largely due to their ease and efficiency of production from a variety of easily existing ingredients and could be cheaper for commercial applications. When reduced to the nanoscale, unique size dependent properties of nanoparticles are manifested [10]. The principal factors believed to cause properties of nanomaterials to differ from their larger micron-sized bulk counterparts include an increase in relative surface area, a greater percentage of atoms at the material's surface, quantum effects which can affect chemical reactivity, and other physical and chemical properties [9,10]. The positioning of the vast majority of nanostructure atoms at the material's surface maximizes their ability to be loaded with therapeutic drugs, and to deliver these agents to target cells and tissues.

The size of nanoparticles, which is comparable to naturally occurring biological molecules, is another feature that makes them well suited for biological applications. Their nanoscale size allows their internalization into cells, and allows them to interact with biomolecules within or on the cell surface, enabling them to potentially affect cellular responses in a dynamic and selective manner. The size of nanoparticles can facilitate their entry into tumor tissues, and their subsequent retention, by a process recognized as the enhanced permeation and retention (EPR) effect. Therapeutic approaches making use of the EPR effect are now recognized as the "gold-standard" in the design of new anti-cancer agents. The EPR phenomena can be described as a combination of "leaky" tumor blood vessels due to alterations in angiogenic regulators, enlarged gap junctions between endothelial cells, and compromised lymphatic drainage in the tumor microenvironment. This localized imbalance allows nanoparticles of certain sizes [11] to readily enter, but to be passively retained within the tumor interstitial space, thereby improving therapeutic potential. In a recent report,particles of 100–200 nm size showed a 4-fold higher rate of tumor uptake compared to particles greater than 300 nm, or less than 50 nm in size [12].

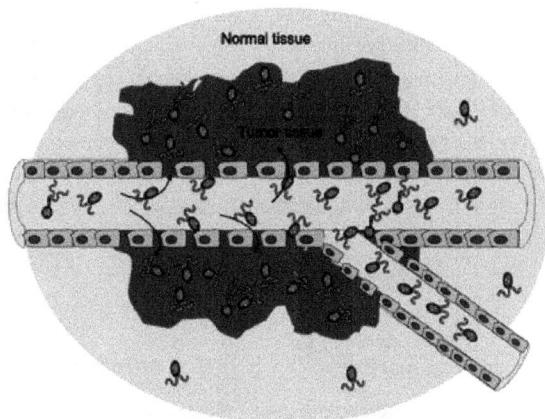

Figure 2. Tumor targeting of nanoparticles passively by enhanced permeability and retention. Long-circulating therapeutic nanoparticles accumulate passively in solid tumor tissue by the enhanced permeability and retention effect. Angiogenic tumor vessels are disorganized and leaky. Hyperpermeable angiogenic tumor vasculature allows preferential extravasation of circulating nanoparticles.ref. 11.

Although smaller nanoparticles don't readily make use of the EPR/enhanced permeation and retention effect, they typically exhibit more nanotoxicity related to their larger surface area/volume ratio [11,12]. These seemingly conflicting actions with respect to nanoparticle size and anti-tumor activity can make it difficult to reliably predict nanoparticle characteristics likely to provide the best therapeutic efficacy without direct testing. The electrostatic nature of nanoparticles is another important consideration as electrostatic interactions between positively charged nanomaterials and target cells are believed to play an important part in cellular adhesion and uptake [13]. Compared to normal eukaryotic cells whose outer leaflet consists of neutral charged zwitterionic phospholipids [14], cancer cells frequently maintain a high concentration of anionic phospholipids on their outer leaflet and large membrane potentials [15–17], and over-express specific groups of charged proteins and carbohydrates [8]. In addition, studies have shown that intracellular pH increases with cell cycle progression and proliferation [18,19], which could affect electrostatically-driven interactions with charged particles at the cell membrane. Even more compelling is data demonstrating that while polycationic polymer particles and cationic fullerenes cause substantial disruption of biomembranes, their neutral or negatively charged counterparts fail to cause measurable effect [20]. While nanoparticles with higher positive charge may be desirable for higher toxicity to cancer cells, very high positive charge may not be suitable for *in vivo* cancer treatment due to rapid serum clearance [21]. Thus, tailoring the surface charge of nanoparticles is expected to influence their cytotoxicity and will likely be an important parameter for developing cancer therapies. The overall shape and morphology of the nanomaterial is another important consideration. Since nanoparticles can be also minute and subtle semiconductors with quantization properties, they can participate in cellular redox reactions and have photocatalytic activity, they are increasingly being considered for use in biomedical applications.

3. Use of metal oxide nanoparticles in early cancer detection

The biomedical applications of metal oxide and ZnO nanomaterials under development at the experimental, preclinical, and clinical levels. A discussion regarding the advantages, approaches, and limitations surrounding the use of metal oxide nanoparticles for cancer applications and drug delivery is presented. The scope of this paragraph is focused on metal oxide nanomaterial systems, and their proposed mechanisms of cytotoxic action, as well as current approaches to improve their targeting and cytotoxicity against cancer cells. Although nanoparticles of many different types of materials can be produced, compatibility issues with living cells limits the types of nanomaterials under consideration for use in biomedical applications. ZnO is considered to be a "GRAS" (generally recognized as safe) substance by the FDA. However, the GRAS designation most commonly refers to materials in the micron to larger size range, as even these substances when reduced to the nanoscale can develop new actions of toxicity. As a result, a detailed evaluation of nanomaterials toxicity in both *in vitro* and *in vivo* systems is needed, as well as identifying means to reduce unwanted toxicity. One common approach to increase biocompatibility and reduce particle aggregation involves coating nanoparticles with discrete sized polymers to render them less toxic, more likely to be taken up by cells, and potentially more suitable for drug delivery applications [22,23].

Interest is growing regarding the use of ZnO and other metal oxide nanomaterials for use as biomarkers for cancer diagnosis, screening, and imaging. Recent studies have shown that ZnO nanoparticle cores capped with polymethyl methacrylate are useful in the detection of low abundant biomarkers [24]. These nanobeads work by facilitating surface absorption of peptide/proteins from cell extracts enabling increased sensitivity and accuracy of cancer biomarker detection using mass spectrometry. Using another approach, a ZnO nanorod based cancer biomarker assay has been developed for high-throughput detection of ultralow levels of the telomerase activity for cancer diagnosis and screening [25]. In an additional approach, multiple reports have described the successful use of iron oxide nanoparticles as contrast agents for cancer detection. Super-paramagnetic oxide nanoparticles coated with a cell resistant polymer have been shown to accumulate within tumor sites via the EPR/enhanced permeation and retention effect in tumor xenograft mice model using magnetic resonance imaging [26]. In another report, the surface of nanoparticles composed of an iron oxide core and oleic acid coating were modified with various pluronic and tetronic block copolymers and shown to provide superior *in-vivo* tumor imaging properties compared to Feridex IV, a commonly used contrast agent [27]. These modified nanoparticles exhibited an extended systemic circulation half-life and reduced clearance properties allowing them to diffuse throughout the tumor vasculature to act as whole tumor contrast agents. While the super-paramagnetic properties of iron oxide nanoparticles offer an advantage for magnetic resonance imaging compared to ZnO, ZnO composite nanomaterials may ultimately prove useful for tumor imaging in the future [28].

Folic acid being a multi dentate ligand, helps in controlling ZnO nanoparticle size through its surface charge density . Also, folic acid has a natural affinity towards Folate receptor protein, which is over expressed by a number of tumor cells. Since ZnO nanoparticles are cytotoxic and can combat the growth of tumor cells, it is envisaged that such a capping would help in targeting tumor cells.

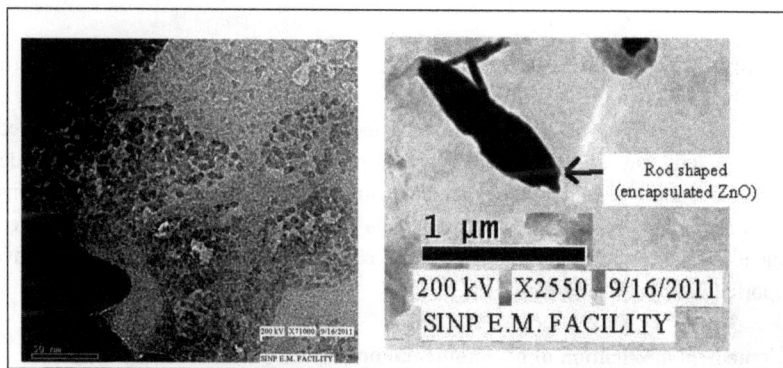

Figure 3. a) ZnO nano –material ~7nm and b) large rod shaped structure of encapsulated Zno in folic acid molecules.

In a similar way gallium-oxy-hydroxide GaO(OH) nanoparticles grown under physiological conditions have proven useful[29], the same was surface conjugated with a gaint sugar molecule and its cytotoxicity effects was useful [20] to combat the survival of cancerous HeLa cells, which was tested *in-vitro*.

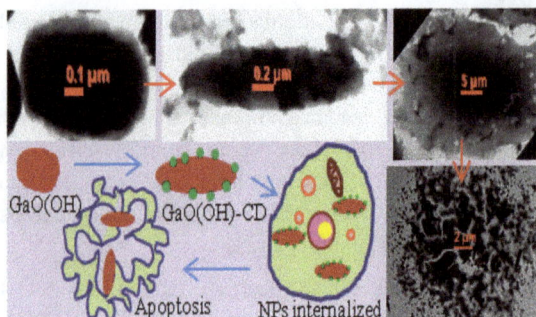

Figure 4. Over all demonstration of gallium-oxy-hydroxyl nanoparticles,then conjugation with cyclodextrin (spindle shape),further internalization into HeLa cells and then apoptosis of the cells (ref.29. copyright American Chemical society)

In a similar way,bio-active nanomaterials, namely: ruthenium hydrous oxide (or ruthenium oxy-hydroxide), $RuO_x(OH)_y$ and also a surface-conjugated novel material of the same within the template of an amino acid molecule: L-cysteine have been studied. These compounds have been prepared through a simple wet chemical route, under physiological condition, such that they could be suitably used in anti-cancer applications. The hydrodynamic size of the prepared NPs were measured. Further, biological consequences of these NPs on cancerous HeLa cells and their cytotoxicity effects have been reported with MTT assay[30].

4.   Biomedical application of plasmonic nanoparticles :

A more novel group in terms of their use in biomedicine are plasmonic nanoparticles, which offer many advantages in biomedical research due to their unique feature, that is, displaying localized surface plasmon resonance (LSPR) bands in the UV–visible-near IR spectral range (for a review on the subject see e.g. [31]). The LSPR frequency is extremely sensitive to subtle changes in the physicochemical environment, for example, the distance between nanoparticles [32], and is also characteristic of their size and shape. Remarkably, in some cases the associated plasmon shift is so dramatic that a color change can be read out by the naked eye and does not require expensive or sophisticated instrumentation [33]. Plasmonic nanoparticles are mainly based on Au or Ag cores because a large array of synthetic procedures exist to produce them in different shapes and sizes, as well as for their outstanding plasmonic properties. The surface chemistry of plasmonic nanoparticles is well-known and widely advanced, thus biofunctionalization can be effectively performed [34]. The characteristic LSPR band of gold (Au) and silver (Ag) nanoparticles can be used for sensing, to trigger light-based events or even for both actions simultaneously. Indeed, their optooelectrical properties render plasmonic nanoparticles a most relevant building block in functional materials, while the fabrication and characterization procedures of such materials are currently changing according to the requirements of biomedicine [35–37]. Probably one of the techniques with a most remarkably growing interest and development is surface enhanced Raman scattering (SERS) spectroscopy [38]. SERS is a label free, highly specific technique, which can also be highly sensitive since water does not interfere with Raman scattering signals whereas most biomolecules are Raman-active. SERS can be easily implemented within optical microscopes, and the use of NIR radiation as an excitation source allows a long penetration depth into biological tissues. One of the challenging issues in using SERS with cellular systems is how to design nanoparticles that are successfully internalized by living cells and yet keeping the integrity of the SERS signal detectable and informative. Cellular composition and their ability to uptake nanoparticles vary widely depending on the cell type and the tissue from which it is retrieved. It should also be noted that the intracellular environment is highly diverse, featuring a wide range of organelles (e.g. nucleus, mitochondira, lysosomes) and soluble factors which can have large effects on the ability to distinguish a SERS signal from the background.

In the illustrated example , the integration of plasmonic AuNPs and carbon nanotubes onto silica microparticles allowed Skirtach et al. to design a tailored microparticle that could be used as a SERS platform to map fibroblast cells. Due to the proximity or 'hot spots' that arise between the AuNPs and carbon nanotubes, SERS spectra corresponding to intracellular biomolecules were successfully detected (Fig. 5) [39]. This strategy is interesting in that the building blocks were chosen taking into account both functionality and biocompatibility, and the coupling of the nanocarbon structures and plasmonic nanoparticles is known to offer large enhancements in SERS. It should be noted that for effective mapping to take place, cells were electroporated, to allow maximum particle uptake to all possible organelles in the cell. The nucleus and cytoplasm could be clearly distinguished by their Raman signatures, specifically aromatic ring stretching in nucleic acids and C–H stretching from proteins and lipids, respectively. These are very complex SERS spectra, requiring powerful data treatment, however the example shows how cells can be successfully mapped using label-free SERS. The second approach describes a SERS mapping method in which signals from extracellular nanoparticles can be disregarded. In studies of nanoparticle uptake, extracellular nanoparticles are typically removed by washing. However, the possibility remains that NPs are bound to the membrane of cells, and given the poor resolution of SERS microscopy techniques in the z plane, recorded signals may arise from both intra- and extra-cellular NPs. To overcome this limitation Xie et al. incubated plasmonic Au nanostars functionalized with Alexa Fluor 750, a widely used dye for SERS with HeLa cervical cancer cells, well known for exhibiting high levels of uptake. Incorporation of a highly effective quencher for the fluorescence and Raman signal of Alexa-750 that cannot cross the cell membrane (tris(2-carboxyethyl)phosphine) guaranteed that exclusively the SERS signal from the Au nanostars@Alexa- 750 laying outside the living cells was quenched. Imaging the living cell culture with Raman microscopy, the distribution of nanoparticles inside the living cell could be registered (Fig. 5) [40].

Nanomedicine-based hyperthermia is another promising therapy for cancer treatment. Infusing a tumor with magnetic or metal nanoparticles, and then exposing the patient to an alternating magnetic field or shortwave radiofrequency, energy produces heat which warms areas immediately adjacent to the nanoparticles [41,42]. When sufficient supernormal temperatures are reached, the tumor cells are killed without harming surrounding healthy tissue. Both photodynamic and hyperthermic noparticle-based cancer approaches share the challenge of preferentially accumulating at tumor sites, unless targeting strategies are also employed.

5.   Immunoassay with nanoparticles

Combination of plasmonic nanoparticles with such immunoassay systems has been quoted as plasmonic ELISAs(enzyme-linked immunosorbant assays (ELISAs). Plasmonic ELISAs have opened up new possibilities for detection of pathogens at low LOD and for detection 'in the field' where complex detection systems are often inapplicable. In practice, detecting with a sectrophotometer or a similar detector, the transition between aggregated and non-aggregated nanoparticles results in a visible colour change that can be read by eye, giving a simple yes or no result. Several examples of plasmonic immunoassays with visible read-outs have been developed for HIV detection [43-45] . In the work by Rica & Stevens , the capsid HIV-1 antigen p24 was detected in human plasma down to a concentration of 10^{-18} g mL $^{-1}$ [43], a viral load that is undetectable by other standard tests. In the same way many other infectious disease could be assayed with convenience [46-48].

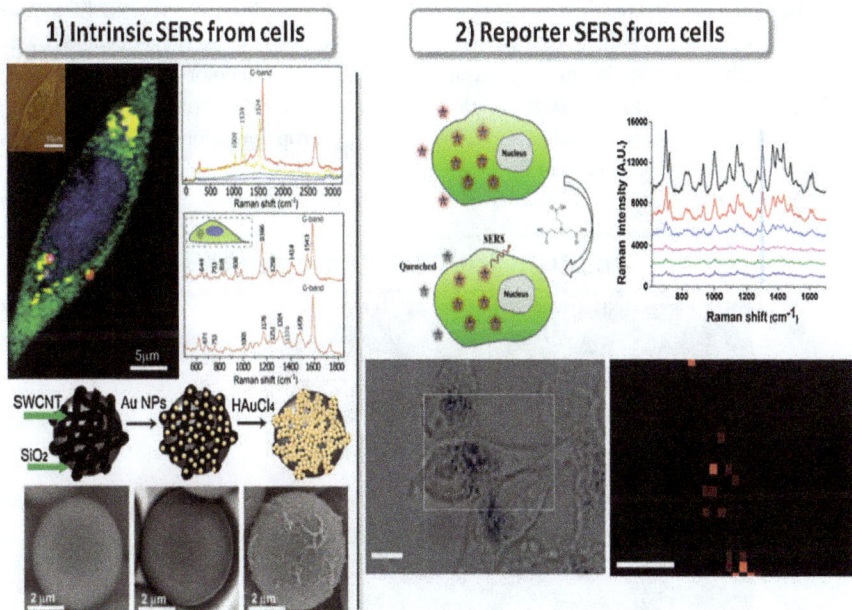

Figure 5. SERS imaging shows the intracellular space of living cells. (1) A silica microparticle containing SWCNT and Au nanoparticles can efficiently enhance the Raman signal of intracellular components of a living cell. Nucleus and cytosol regions are mapped by SERS, providing a detailed picture of the inner structure of the cell [39]. Copyright 2013, Wiley-VCH. (2) Plasmonic nanoparticles functionalized with a Raman reporter that is selectively quenched outside the living cell yield a SERS signal acquired exclusively from nanoparticles internalized by the living cell [40]. Copyright 2014, Royal Society of Chemistry.

6. Diagnostic study of tumor by radioactive imaging

Tumors are cell clusters which lack a normal blood supply, delivery of pharmaceuticals via nanoparticles to the tumor cells, either for detection or treatment purposes, is a key issue [49,50]. Cancer cell targeting ligands are usually attached to the nanoparticles surface so that localized targeting can be carried out while limiting damage to surrounding healthy cells. The use of radio-pharmaceutics with targeted nanoparticles has been a key advancement to cancer diagnosis and subsequent monitoring, however often translation into *in vivo* settings is lengthy and achieves limited success. In a recent development, a core–shell nanoparticle composed of Au and silica has been shown to successfully target melanoma by exploiting the cellular receptors for an amino acid RGDY sequence [51]. The nanoparticles also contain a fluorophore (Cy5) and a radiolabel (^{124}I), thereby allowing optical and positron emission tomography (PET) imaging respectively as well as simultaneously.

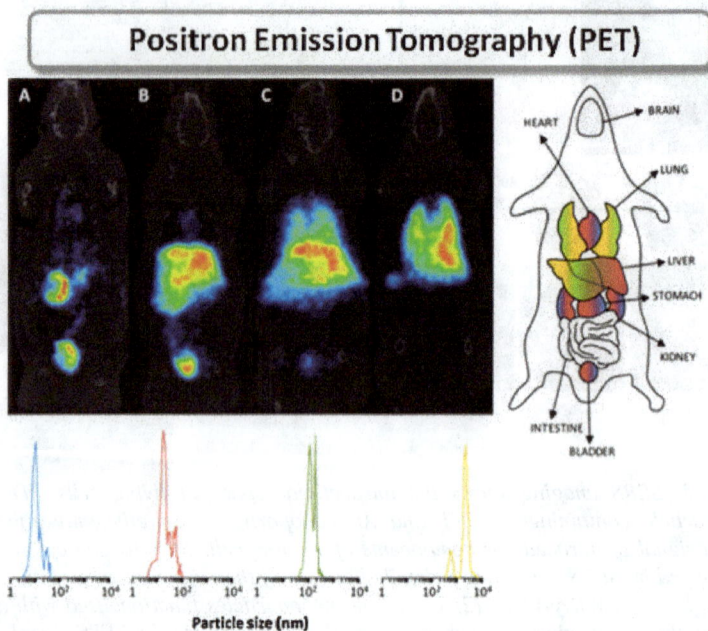

Figure 6. Detecting nanoparticle bio-distribution in animal models. Positron emission tomography (PET) allows the in vivo analysis of the bio-distribution of radioactively labelled aluminium oxide nanoparticles. This procedure can be extended to any given nanoparticle by using a suitable radioactive activation protocol. Copyright 2013, American Chemical Society.

PET for studying nanoparticle distribution is based on recording the emission of two gamma quanta by positrons associated with radioactive isotopes labelled within the nanoparticles under study [52,53]. This technique offers the advantages of a definitely large sensitivity and space-resolving capability. Moreover, provided that an adequate radio-activation reaction is available, PET can in principle be used to follow the biodistribution of any type of nanoparticles. Llop et al. demonstrated the different biodistribution of aluminum oxide nanoparticles, as a function of core size, in an elegant application of PET imaging (Fig. 6) [54]. By radioactive labelling of oxygen   atoms in the aluminum oxide nanoparticles, the bioaccumulation of the nanoparticles in different organs could be studied *in vivo*. As noted in that study, the decay time of the radioactive isotope is relatively short, in the range of 1 hour. However, radioactive   labelling cannot be used for bio-distribution studies comprising longer periods of time such as days or weeks, for example for dynamic studies on the biological fate of nanoparticles. Also, an easy renal clearance of the compound (drug) is a must, as the radio-pharmaceutics may accumulate ,which is highly undesirable. We anyway highlight this approach as this limitation can be overcome by using a suitable radiolabel, while the concept can be readily translated to other nanoparticles such as gallium hydrous oxide or ruthenium hydrous oxide as described in the references [29] and[30] in future as the medical cyclotron production along with research in nano-material design for biomedicine comes in the forefront platform.

7.   Conclusion

The interface between materials science and biomedicine is providing new and exciting avenues of research, however the reality is that issues associated with nano-toxocity and *in vivo* clearance are limiting factors. Moreover each type of nanoparticle has its own characteristics, as also the size parameters and surface properties can vary largely, accordingly their biological interplay could be different. The development of new physical and chemical properties that can accompany reduction of materials to the nanoscale offers advantages for developing anticancer agents, including the ability to tailor the electrostatic properties and size of nanoparticles to promote cellular uptake and make use of the enhance permeation and retention effect to promote intra-tumor accumulation. In this article, the promising features of nanoparticles as diagnostic and drug-delivery systems with biomedical performance has been highlighted. Given their unique optoelectrical properties, plasmonic nanoparticles are at the forefront of these novel opportunities. For advances in the development of hand-held nanoparticle diagnostic devices or in 3D imaging as examples, or in radiological diagnostic studies research at the interface between materials science and biomedicine requires further efforts.

References

[1]   A.P. Alivisatos, The use of nanocrystals in biological detection., Nat. Biotechnol. 22 (2004) 47-52. https://doi.org/10.1038/nbt927

[2]   E. Scott McNeil , Nanotechnology for the biologist., J. Leukoc. Biol. 78(2005) 585–594. https://doi.org/10.1189/jlb.0205074

[3]   N.L.Rosi, C.A. Mirkin, Nanostructures in bio-diagnostics.Chem.Rev.105 (2005), 1547-1562. https://doi.org/10.1021/cr030067f

[4]   M. Ferrari, Cancer nanotechnology: opportunities and challenges. Nat Rev. Canncer 5(2005),161–171. https://doi.org/10.1038/nrc1566

[5]   V. Wagner, A. Dullaart, A.K. Bock, A. Zweck. The emerging nanomedicine landscape. Nat Biotechnol. 24(2006) 1211-1217. https://doi.org/10.1038/nbt1006-1211

[6]   Y.X.J. Wang, S.M. Hussain, G.P. Krestin, Superparamagnetic iron oxide contrast agents: physicochemical characteristics and applications in MR imaging. Eur. Radiol. 11 (2001), 2319. https://doi.org/10.1007/s003300100908

[7]   Jae-Hyun Lee , Yong-Min Huh, Young-wook Jun, et al., Artificially engineered magnetic nanoparticles for ultra-sensitive molecular imaging. Nat Med 13(2007), 95-99. https://doi.org/10.1038/nm1467

[8]   R.G. Panchal, Novel therapeutic strategies to selectively kill cancer cells. Biochem Pharmacol 55 (1998), 247–252. https://doi.org/10.1016/S0006-2952(97)00240-2

[9]   A Nel, T Xia, L .Madler, N Li., Toxic potential of materials at the nanolevel. Science 311(2006), 622–627. https://doi.org/10.1126/science.1114397

[10]   S Lanone, J. Boczkowski,. Biomedical applications and potential health risks of nanomaterials: molecular mechanisms. Curr Mol Med 6(2006), 651–663. https://doi.org/10.2174/156652406778195026

[11]   K .Cho, X .Wang, S. Nie, et al. Therapeutic nanoparticles for drug delivery in cancer. Clin Cancer Res. 14 (2008),1310–1316. https://doi.org/10.1158/1078-0432.CCR-07-1441

[12]   Li SD, Huang L. Pharmacokinetics and biodistribution of nanoparticles. Mol Pharm 5(2008) 496-504. https://doi.org/10.1021/mp800049w

[13] M Ohgaki, T.Kizuki, M. Katsura, K Yamashita, Manipulation of selective cell adhesion and growth by surface charges of electrically polarized hydroxyapatite J Biomed Mater Res. 57(2001), 366–373. https://doi.org/10.1002/1097-4636(20011205)57:3<366::AID-JBM1179>3.0.CO;2-X

[14] S Hong, A .Mecke, et al. Nanoparticle interaction with biological membranes: does nanotechnology present a Janus face? Acc Chem Res. 40 (2007), 335–342. https://doi.org/10.1021/ar600012y

[15] M.Abercrombie, E.J. Ambrose. The surface properties of cancer cells: a review. Cancer Res. 22 (1962) , 525–548.

[16] JOM Bockris, M.A. Habib. Are there electrochemical aspects of cancer? J Biol Physics 10 (1982), 227–237. https://doi.org/10.1007/BF01991943

[17] N .Papo, M. Shahar, L. Eisenbach, Y. Shai. A novel lytic peptide composed of DL-amino acids selectively kills cancer cells in culture and in mice. J Biol Chem 278 (2003), 21018–21023. https://doi.org/10.1074/jbc.M211204200

[18] LD Shrode, H. Tapper, S.Grinstein Role of intracellular pH in proliferation, transformation, and apoptosis. J Bioenerg Biomembr .29 (1997), 393–399. https://doi.org/10.1023/A:1022407116339

[19] Rich IN, Worthington-White D, Garden OA, Musk P. Apoptosis of leukemic cells accompanies reduction in intracellular pH after targeted inhibition of the Na(+)/H(+) exchanger. Blood 95 (2000), 1427–1434.

[20] Y.J .Tang, J.M .Ashcroft, D. Chen, et al. Charge-associated effects of fullerene derivatives on microbial structural integrity and central metabolism. Nano Lett 7 (2007), 754–760. https://doi.org/10.1021/nl063020t

[21] P. Xu , E.A. Van Kirk, Y Zhan, et al. Targeted charge-reversal nanoparticles for nuclear drug delivery. Angew Chem Int Ed Engl. 46(2007), 4999–5002. https://doi.org/10.1002/anie.200605254

[22] UO Hafeli, JS Riffle, L Harris-Shekhawat, et al. Cell uptake and *in vitro* toxicity of magnetic nanoparticles suitable for drug delivery. Mol Pharm 6 (2009),1417–28. https://doi.org/10.1021/mp900083m

[23] WI Hagens, AG Oomen, WH de Jong, et al. What do we (need to) know about the kinetic properties of nanoparticles in the body? Regul Toxicol Pharmacol 49(2007), 217–229. https://doi.org/10.1016/j.yrtph.2007.07.006

[24] W. Shen, H. Xiong, Y. Xu, et al. ZnO-poly(methyl methacrylate) nanobeads for enriching and desalting low-abundant proteins followed by directly MALDI-TOF

MS analysis. Anal Chem. 80 (2008), 6758–6763.
https://doi.org/10.1021/ac801001b

[25]   A Dorfman, O Parajuli, N Kumar, J.I. Hahm Novel telomeric repeat elongation
       assay performed on zinc oxide nanorod array supports. J Nanosci Nanotechnol
       8(2008), 410–415. https://doi.org/10.1166/jnn.2008.146

[26]   H. Lee, E .Lee, K .Kim do, et al. Antibiofouling polymer-coated
       superparamagnetic iron oxide nanoparticles as potential magnetic resonance
       contrast agents for *in vivo* cancer imaging. J Am Chem Soc.128(2006), 7383–
       7389. https://doi.org/10.1021/ja061529k

[27]   TK Jain, J Richey, M.Strand, et al. Magnetic nanoparticles with dual functional
       properties: drug delivery and magnetic resonance imaging.
       Biomater.29(2008),4012–4021. https://doi.org/10.1016/j.biomaterials.2008.07.004

[28]   Sreetama Dutta and Bichitra N Ganguly. . Characterization of ZnO nano particles
       grown in presence of Folic Acid template. J. Nanobiotechnology 10 (2012),29
       10pages.

[29]   Bichitra Nandi Ganguly, Vivek Verma, Debanuj Chatterjee, Biswarup Satpati,
       Sushanta Debnath and Partha Saha, . Study of Gallium Oxide Nanoparticles
       Conjugated with β-cyclodextrin -An Application to Combat Cancer, ACS
       Materials and Interfaces : 8, (2016), 17127- 17131.
       https://doi.org/10.1021/acsami.6b04807

[30]   Bichitra Nandi Ganguly , Buddhadeb Maity, Tapan Kumar Maity, Joydeb
       Manna, Modhusudan Roy, Manabendra Mukherjee , Sushanta Debnath, Partha
       Saha, Nagaraju Shilpa, and Rohit Kumar Rana, . l-Cysteine-Conjugated
       Ruthenium Hydrous Oxide Nanomaterials with Anticancer Active Application,
       Langmuir 4 (2018),1447-1456. https://doi.org/10.1021/acs.langmuir.7b01408

[31]   M.A. Garcia, Surface plasmons in metallic nanoparticles: fundamentals and
       applications J. Phys. D: Appl. Phys. 44 (2011) , 283001 (20pages).

[32]   P.K. Jain, M.A. El-Sayed, Plasmonic coupling in noble metal nanostructures
       Chem. Phys. Lett. 487 (2010), 153-164.
       https://doi.org/10.1016/j.cplett.2010.01.062

[33]   Y. Song, W. Wei, X. Qu, Colorimetric biosensing using smart materials, Adv.
       Mater. 23 (2011), 4215-4236. https://doi.org/10.1002/adma.201101853

[34]   R.M. Fratila, Strategies for the biofunctionalization of gold and iron oxide
       nanoparticles. Langmuir 30 (2014), 15057-71. https://doi.org/10.1021/la5015658

[35] P.D. Howes, S. Rana, M.M. Stevens, Plasmonic nanomaterials for biodiagnostics. Chem. Soc. Rev. 43 (2014), 3835-3853. https://doi.org/10.1039/C3CS60346F

[36] N.E. Motl, A. F. Smith, C. J. DeSantis and S. E. Skrabalak, Engineering plasmonic metal colloids through composition and structural design. Chem. Soc. Rev. 43 (2014), 3823-3834. https://doi.org/10.1039/C3CS60347D

[37] S.M. Lee, et al. Drug-loaded gold plasmonic nanoparticles for treatment of multidrug resistance in cancer. Biomaterials 35 (2014), 2272-2282. https://doi.org/10.1016/j.biomaterials.2013.11.068

[38] R.A. Alvarez-Puebla, L.M. Liz-Marza´n, Traps and cages for universal SERS detection. Chem. Soc. Rev. 41 (2012), 43-51. https://doi.org/10.1039/C1CS15155J

[39] A. Yashchenok, Nanoengineered colloidal probes for Raman-based detection of biomolecules inside living cells Small 9 (2013), 351-356. https://doi.org/10.1002/smll.201201494

[40] H. Xie, Yiyang Lin, Manuel Mazo, Ciro Chiappini, Ana Sánchez-Iglesias, Luis M . Liz-Marzán and Molly M. Stevens, Identification of intracellular gold nanoparticles using surface-enhanced Raman scattering,.Nanoscale 6 (2014),12403-12407. https://doi.org/10.1039/C4NR04687K

[41] A Jordan, R Scholz, K. Maier-Hauff, et al. The effect of thermotherapy using magnetic nanoparticles on rat malignant glioma. J Neurooncol (2006)78, 7–14. https://doi.org/10.1007/s11060-005-9059-z

[42] A. Jordan, R.Scholz, P .Wust, et al. Magnetic fluid hyperthermia (MFH): Cancer treatment with AC magnetic field induced excitation of biocompatible superparamagnetic nanoparticles. J Magnetism Magnetic Materials (1999) 201, 413–419. https://doi.org/10.1016/S0304-8853(99)00088-8

[43] R. de la Rica, M.M. Stevens, Plasmonic ELISA for the ultrasensitive detection of disease biomarkers with the naked eye. Nat. Nanotechnol. 7 (2012), 821-824. https://doi.org/10.1038/nnano.2012.186

[44] C.D. Chin, Microfluidics-based diagnostics of infectious diseases in the developing world. Nat. Med. 17 (2011) , 1015-1019. https://doi.org/10.1038/nm.2408

[45] W. Qu, Copper-mediated amplification allows readout of immunoassays by the naked eye.Angew. Chem. Int. Ed. 50 (2011), 3442-3445. https://doi.org/10.1002/anie.201006025

[46] Y. Xianyu, Z. Wang, X. Jiang, A Plasmonic Nanosensor for Immunoassay *via* Enzyme-Triggered Click Chemistry. ACS Nano 8 (2014), 12741-1247. https://doi.org/10.1021/nn505857g

[47] X.M. Nie, Huang R, Dong C-X, Tang L-J, Gui R, Jiang J-H Plasmonic ELISA for the ultrasensitive detection of Treponema pallidum. Biosens. Bioelectron. 58 (2014) 314-319. https://doi.org/10.1016/j.bios.2014.03.007

[48] M. Coronado-Puchau, Laura Saa, Marek Grzelczak et al., Enzymatic modulation of gold nanorod growth and application to nerve gas detection. Nano Today 8 (2013), 461- 468. https://doi.org/10.1016/j.nantod.2013.08.008

[49] A.S. Thakor, S.S. Gambhir, Nanooncology: the future of cancer diagnosis and therapy. C.A. Cancer, J. Clin. 63 (2013) 395-418. https://doi.org/10.3322/caac.21199

[50] M. Sivasubramanian, Y. Hsia, L.-W. Lo, anoparticle-facilitated functional and molecular imaging for the early detection of cancer. Front. Mol. Biosci. 1 (2014).

[51] E. Phillips, et al. Clinical translation of an ultrasmall inorganic optical-PET imaging nanoparticle probe. Sci. Transl. Med. 6 (2014) 260ra149. TRIAL REGISTRATION: ClinicalTrials.gov NCT01266096.

[52] J. Llop, V. Go´mez-Vallejo, N. Gibson, Quantitative determination of the biodistribution of nanoparticles: could radiolabeling be the answer? Nanomedicine (Lond.) 8 (2013), 1035-1038. https://doi.org/10.2217/nnm.13.91

[53] Bichitra Nandi Ganguly, Nagendra Nath Mondal, Maitreyee Nandy, Frank Roesch, Some Physical Aspects of Positron Annihilation Tomography: a critical review; Journal of Radioanalytical and Nuclear Chemistry 279(2009), 685-698. https://doi.org/10.1007/s10967-007-7256-2

[54] C. Pe´rez-Campan˜a, et al. Biodistribution of Different Sized Nanoparticles Assessed by Positron Emission Tomography: A General Strategy for Direct Activation of Metal Oxide Particles. ACS Nano 7 (2013), 3498-3505. https://doi.org/10.1021/nn400450p

Chapter 9

Conclusive Remarks

In the recent times, synthesis of functionalized nano-scaffolds have demonstrated tremendous potential to greatly enhance the clinical armamentarium for cancer theranostics. Extensive research efforts at the interface between materials science and biomedicine have resulted in exceptional accomplishments toward syntheses of various types of nanoplatforms that can directly be used for biomedical research. Today, increasing numbers of nanoparticle-based diagnostic or therapeutic agents are either being commercialized or have reached the clinical stage, thereby, achieving important milestones in "bench-to-bedside" translation of nanomaterial research and nanotechnology. Nanoparticles exhibit unique size-dependent physical and chemical properties, which if properly harnessed can address unsolved challenges in clinical oncology. Particularly, they possess large functional surface areas, easily controllable surface chemistry that facilitates surface functionalization to achieve tailored characteristics for effective use in personalized disease management. In the endeavour toward translating this promise into clinical reality, visualization of the distribution of nanoparticle-based carriers in the body following systematic administration through any route is of paramount importance. Presently, the most prudent approach that provides quantitative information about the whole body bio-distribution is by incorporating suitable radioisotopes in the nanoparticles — a process known as "radiolabeling". After administration of the radiolabeled nanoparticles in living subjects, their *in vivo* bio-distribution can be non-invasively monitored by molecular imaging techniques such as single photon emission computed tomography (SPECT), PET, Cerenkov luminescence (CL), Cerenkov resonance energy transfer (CRET), etc., that are now been widely explored for cancer imaging in preclinical and/or clinical settings.

In addition to nano-drug carriers, interest is growing regarding the ability of certain nanomaterials to mediate anti-cancer effects on their own, including metal oxides. One approach involves the successful use of metal oxide nanoparticles to kill cancer cells when UV irradiated. In these studies, HeLa cells were completely killed in the presence of TiO2 and UV irradiation, and *in vivo* tumor growth arrested up to 30 days, while no cancer cell killing was observed in the absence of the metal oxide nanoparticles and UV light. Although effective for the treatment of skin cancer, a limitation of this photodynamic nanomedicine-based approach is the inability of UV light to penetrate more than 1 mm through skin, unless fiber optics or surgery are used in conjunction.

Nanomedicine-based hyperthermia is another promising therapy for cancer treatment.

Infusing a tumor with magnetic or metal nanoparticles, and then exposing the patient to an alternating magnetic field or shortwave radiofrequency energy produces heat which warms areas immediately adjacent to the nanoparticles. When sufficient supernormal temperatures are reached, the tumor cells are killed without harming surrounding healthy tissue. Both photodynamic and hyperthermic nanoparticle-based cancer approaches share the challenge of preferentially accumulating at tumor sites, unless targeting strategies are also employed. In addition to the above described applications, emerging approaches using zinc oxide nanoparticles are gaining interest for the development for new anti-cancer therapeutics.

However, there are also limitations in the use of such nanoparticles for *in vivo* application and research as they may accumulate in various organs of the body. Therefore an important strategy would be to have conjugants/ligands which have mainly been used as complexing agent to increase aqueous solubility of poorly soluble drugs and to increase their bio-availability and stability. Since these giant complexing agents possess hydrophilic polar groups projected outward, it renders such compounds water-soluble, keeping the hydrophobic moiety intact. This type of conjugation also enhances the water solubility of the encapsulated drug so that it could be easily taken up by the malignant cells (those are rapidly dividing) and would be appropriate for the renal clearance of the same after its administration and investigation time duration in molecular imaging processes (such as in PET), which is very important and hence qualifies as a drug carrier.

Apart from these, there have been many other application described through the various articles in the chapters of this book. It is envisaged that the information would be helpful for the researchers.

Keyword Index

Anticancer 147

Bio-applications 123
Bio-imaging 19
Biomedicine 179

Debye Scherrer Method 104
Defects 123
Drug Delivery 147, 179

Electron Microscopy 104

Femtosecond Fluorescence
Upconversion 75

Imaging 147

Laser Flash Photolysis 75
Light Scattering 104

Metal/Oxide Nanoparticle 50

Nanoparticle 50
Nanoscale 50

Optical Properties 19

Organometallic Compound 50

Photo induced Electron
Transfer (PET) 75
Phototherapy 147
Plasmon Absorption Band 75
Positron Annihilation Technique 123

Quantum Dots 19

Ru:CNDEDAs 75

Semiconductor Nanocrystals 19
Semiconductors 123
Sensing 147
Surface Functionality 179
Swallowing Surgeons 147
Synthesis 19

TCSPC 75
Therapeutic and
Diagnostic Applications 179
Thermal Decomposition 50

UV-Vis Absorption Spectroscopy 75

X-Ray Diffraction 104

About the Editor

Dr. B. Nandi Ganguly (from Saha Institute of Nuclear Physics, KOLKATA, INDIA), has a long experience of research, as well as teaching in post graduate courses. She also has collaborated with internationally known senior scientists of reputed institutions.

Initially, she worked on radio-chemical separations of metallic species using surface active agents with high selectivity owing to their chemical reactive mechanisms, there by specializing in micellar and reverse-micellar as well as in micro-emulsion processes. Later, her interest grew up on studying the subtleties of molecular association, through various techniques. But mainly thereafter, her continued interest was sustained on using positron annihilation techniques as chemical probe which led to studies of various kinds of intriguing molecular processes in liquids. Thereafter, her interest continued on studying various kinds of nano-porous materials, polymers, hydrogen bonded supra-structures, as well as nanomaterials using various state-of-the-art techniques and has ventured into the myriads of interdisciplinary research. Of late, her interest has been focused on studying conjugated nano-metal oxide particles under physiological conditions, for utilization in bio-medical diagnostic as well as therapeutic processes thus fostering and developing anti- cancer research applications.